MW00473870

ELECTROMAGNETIC PRINCIPLES OF INTEGRATED OPTICS

ELECTROMAGNETIC PRINCIPLES OF INTEGRATED OPTICS

DONALD L. LEE
Raytheon Research Division

JOHN WILEY & SONS

NEW YORK   CHICHESTER   BRISBANE   TORONTO   SINGAPORE

Library of Congress Cataloging-in-Publication Data:

Lee, Donald L.
 Electromagnetic principles of integrated optics.

 Includes index.
 1. Integrated optics.   2. Electromagnetic theory.
I. Title.

TA1660.L44   1986   621.36′93   86-4042
ISBN 0-471-87978-9

Printed in the United States of America

10 9 8 7 6 5 4 3 2 1

To the memory of my father,
Walter Lee

About the Author

Donald L. Lee is a Senior Scientist with the Raytheon Research Division in Lexington, Massachusetts. He is presently working on the development of advanced monolithic electro-optical circuits for infrared signal detection. His previous responsibilities included the analysis and design of new types of acoustic-wave devices for signal processing and oscillator applications. While a faculty member of the Electrical Engineering Department and Laboratory for Surface Science and Technology at the University of Maine, he conducted research in the area of microwave acoustics, initiated a new course on integrated optics, and taught courses in acoustics and electromagnetics. He obtained his M.S. and Ph.D. degree in Electrical Engineering from the Massachusetts Institute of Technology.

Dr. Lee is the author of numerous papers, holds several patents, and was the 1980 recipient of the IEEE group on Sonics and Ultrasonics award for best paper of the year. He is a member of the IEEE.

Preface

The purpose of this text is to provide an alternative to conventional presentations of time-harmonic electromagnetics at the undergraduate level. The emphasis is on principles, phenomena, and methods appropriate to optical rather than microwave frequency spectra. This approach is motivated by the significant growth in the areas of optical communication and signal processing within the last decade. The emerging area of integrated optics is used as the focus of the text because it introduces in a unified fashion not only a large number of the existing guided-wave optical components but also many of the physical and mathematical concepts that are fundamental to the understanding of electromagnetics at optical frequencies. Because the field of integrated optics is still young and its ultimate direction not yet well defined, emphasis is placed upon the principles behind device operation rather than engineering details. Therefore, in many cases simplified component geometries are chosen that emphasize concept clarity rather than commercial practicality.

Unlike many other texts written on the subject of integrated optics, this book is intended to be a teaching rather than a reference work. It is, therefore, completely self-contained. All required results are derived as needed and lengthy, complicated appendices are avoided. More difficult or advanced topics not essential to an overall understanding of the text are provided at the end of several chapters. It is assumed that the student has a mathematical background and a comprehension of Maxwell's equations that is consistent with the completion of a standard junior level course in electrostatics. No prior knowledge of wave propagation beyond the introductory physics level is required.

The material in this text may be presented in several different fashions. For curricula in which electromagnetics is taught in one term, it may be viewed as a follow-on elective that treats the optical aspects of electromagnetics within the context of integrated optics. For more in-depth two-term sequences, divided roughly along the lines of time-independent and -dependent solutions to Maxwell's equations, this text provides for an alternative presentation of the second term material oriented toward optical rather than microwave frequencies. Appropriate coverage for either of these teaching formats would consist of Chapters 1–4, 7, and portions of 5, 6, and 8. For students having a substantial background in conventional wave theory, the text can be used in a one-term senior or first-year graduate-level course by omitting the introductory Chapter 1 and the field fundamentals covered in Chapter 2, and substituting the material on Bragg scattering and optical fibers found in Chapters 9 and 10, respectively. Additionally, some of the more advanced topics in Chapters 3–6 and 8 could be included.

This text is an outgrowth of notes developed for a course on integrated optics taught as a senior elective at the University of Maine. It emerged over a four-year period out of the belief that there existed a need for a self-contained, unified treatment of electromagnetics oriented toward guided-wave optics that could be presented at the senior or first-year graduate level. The course has been met with a good deal of student enthusiasm over the years and the comments and criticism from the juniors, seniors, and graduate students who have taken it played an important role in the development of the text.

To give full recognition to the various contributors is an extremely difficult task due to the great volume of literature in existence. Generally, reference is given to only those articles with which the author has greatest familiarity. The author extends his sincere appreciation to Mrs. Barbara Deshane, Mrs. Susan Niles, and Mrs. Janice Gomm for their expert typing of the manuscript, and to his wife, Chris, for her constant encouragement during the preparation of the manuscript.

DONALD L. LEE

Lexington, Massachusetts

Contents

1. INTRODUCTION   **1**

1.1. HISTORICAL DEVELOPMENT OF LIGHTWAVE
TECHNOLOGY   1

1.2. INTEGRATED-OPTICS COMPONENTS   2

    1.2.1. Semiconductor Lasers and Detectors   3

    1.2.2. Dielectric Waveguides   4

    1.2.3. Modulators and Interferometers   5

    1.2.4. Directional Couplers   6

    1.2.5. Gratings and Filters   7

    1.2.6. Integrated-Optics Spectrum Analyzer   7

1.3. ORGANIZATION OF THIS BOOK   9

2. BASIC ELECTROMAGNETIC THEORY   **13**

2.1. MAXWELL'S EQUATIONS   13

2.2. CONSTITUTIVE RELATIONS FOR AN ISOTROPIC
MEDIUM   14

2.3. TIME-HARMONIC SOLUTION OF MAXWELL'S
EQUATIONS   14

2.4. THE WAVE EQUATION IN SOURCE-FREE
ISOTROPIC MEDIA   18

    2.4.1. The Wave Equation   18

    2.4.2. The Dispersion Relation   19

    2.4.3. The Wave Vector **k**   20

    2.4.4. Maxwell's Equations for Plane-Wave Solutions   22

2.5. PHASE VELOCITY   24

2.6. GROUP VELOCITY   26

2.7. POLARIZATION   31

2.8. DUALITY   33

2.9. PLANE-WAVE SPECTRUM OF FINITE-WIDTH BEAMS   36

2.10. POWER FLOW   39
    2.10.1. Poynting's Theorem   39
    2.10.2. Complex Poynting's Theorem   44

3. REFLECTION AND TRANSMISSION AT A DIELECTRIC INTERFACE   **51**

3.1. INTRODUCTION   51

3.2. BOUNDARY CONDITIONS BETWEEN TWO DIELECTRIC INTERFACES   51

3.3. PHASE MATCHING AND SNELL'S LAW   52

3.4. REFLECTION AND TRANSMISSION COEFFICIENTS   58
    3.4.1. TE Wave Incidence   58
    3.4.2. TM Wave Incidence   61

3.5. TOTAL INTERNAL REFLECTION   63

3.6. GOOS–HAENCHEN SHIFT (ADVANCED TOPIC)   67

4. THE SLAB DIELECTRIC WAVEGUIDE   **77**

4.1. INTRODUCTION   77

4.2. THE SYMMETRIC SLAB DIELECTRIC WAVEGUIDE   78
    4.2.1. TE Solutions   79
    4.2.2. Dispersion Relation (ω vs k_z)   85

4.3. THE ASYMMETRIC SLAB WAVEGUIDE   86
    4.3.1. TE Modes   87
    4.3.2. TM Modes   89
    4.3.3. Ray Interpretation of the Guidance Condition   90
    4.3.4. Dispersion Relation for the Asymmetric Slab Waveguide   92
    4.3.5. Guided-Mode Poynting Power   94

4.4. RADIATION MODES   97
    4.4.1. Radiation Modes on the Symmetric Slab Waveguide   97
    4.4.2. Extension to the Asymmetric Slab Waveguide (Advanced Topic)   102

5. PRACTICAL WAVEGUIDING GEOMETRIES   113

5.1. INTRODUCTION   113

5.2. GRADED-INDEX TWO-DIMENSIONAL WAVEGUIDES:
      THE WKB METHOD   116
      5.2.1. Formulation   116
      5.2.2. Guidance by a Parabolic Variation in Permittivity   123
      5.2.3. Ray Trajectory   126

5.3. THREE-DIMENSIONAL WAVEGUIDING   127
      5.3.1. Effective Index Method for the Embedded-Strip
              Waveguide   128
      5.3.2. Extension to Other Waveguide Geometries   134

5.4. RADIATION BENDING LOSSES   135

6. THE PRISM COUPLER   147

6.1. INTRODUCTION   147

6.2. FRUSTRATED TOTAL INTERNAL REFLECTION   147

6.3. WAVEGUIDE EXCITATION USING THE
      PRISM COUPLER   154

6.4. FILM CHARACTERIZATION USING THE
      PRISM COUPLER   166

7. WAVEGUIDE FABRICATION   171

7.1. INTRODUCTION   171

7.2. FABRICATION OF STEP-INDEX DIELECTRIC
      WAVEGUIDES   171
      7.2.1. Sputtered-Film Waveguides   171
      7.2.2. Polymer-Film Waveguides   174

7.3. FABRICATION OF GRADED-INDEX DIELECTRIC
      WAVEGUIDES   176
      7.3.1. Ion-Migration Waveguides   176
      7.3.2. Proton-Exchange Waveguides   178
      7.3.3. Waveguides Formed by Metal In-Diffusion   180
      7.3.4. Ion-Implanted Waveguides   181

7.4.   STEP-INDEX SEMICONDUCTING WAVEGUIDES   185

    7.4.1.   Liquid-Phase Epitaxial Film Waveguides   186

    7.4.2.   Chemical Vapor Deposition Epitaxial Waveguides   189

    7.4.3.   Molecular-Beam Epitaxial Waveguides   192

7.5.   THREE-DIMENSIONAL WAVEGUIDES   194

    7.5.1.   Raised-Channel Waveguides Formed by
          Mask and Etch   194

    7.5.2.   Raised-Channel Waveguides Formed by
          Shadow Masking   196

    7.5.3.   Buried-Channel Waveguides   198

7.6.   OPTICAL-FIBER WAVEGUIDES FORMED BY FLAME
       HYDROLYSIS   199

8.   MODE COUPLING   209

8.1.   INTRODUCTION   209

8.2.   WAVEGUIDE SYMMETRY PROPERTIES   209

    8.2.1. z-Reversal Symmetry   209

    8.2.2.   Time-Reversal Symmetry   211

8.3.   LORENTZ RECIPROCITY THEOREM   212

    8.3.1.   Formulation   212

8.4.   MODE ORTHOGONALITY RELATIONS   215

    8.4.1.   Orthogonality Relations for Dielectric Waveguides   216

8.5.   COUPLED EQUATIONS OF MOTION   218

8.6.   COUPLED WAVEGUIDES   220

8.7.   FORWARD COUPLING: THE DIRECTIONAL COUPLER   224

8.8.   BACKWARD COUPLING: THE GRATING REFLECTOR   231

9.   BRAGG SCATTERING   247

9.1.   THE BRAGG CONDITION   247

9.2.   THE BULK-WAVE ACOUSTO-OPTIC BEAM
       DEFLECTOR   251

9.3.   THE BULK-WAVE ACOUSTO-OPTIC SPECTRUM
       ANALYZER   254

9.4.	GUIDED-WAVE BEAM DEFLECTOR	257
	9.4.1. Deflector Geometry	257
	9.4.2. Coupled-Mode Equations	258
	9.4.3. The Photoelastic Effect	263
	9.4.4. Bragg Deflection Efficiency	265

10. OPTICAL FIBERS   **275**

10.1.	INTRODUCTION	275
10.2.	TYPES OF FIBER GEOMETRIES	275
10.3.	MONOMODE FIBERS	278
	10.3.1. Wave Equation for Weakly Guided Fibers	278
	10.3.2. Stationary Formulation	279
	10.3.3. Solution to the Step-Index Profile	283
	10.3.4. Spatial Distribution of Light Intensity	287
10.4.	MULTIMODE STEP-INDEX FIBER (ADVANCED SECTION)	288
	10.4.1. Wave Equation	289
	10.4.2. Boundary Conditions and the Guidance Condition	293
	10.4.3. Cutoff	295
	10.4.4. Power Density Distribution	298
10.5.	LOSS MECHANISMS IN OPTICAL FIBERS	298
	10.5.1. Absorption	298
	10.5.2. Scattering	299
	10.5.3. Radiation Bending Losses	299
10.6.	SIGNAL DISTORTION DUE TO DISPERSION	301
	10.6.1. Effect of a Dispersive System on a Gaussian Pulse	302
	10.6.2. Material Dispersion	307
	10.6.3. Waveguide Dispersion	309
	10.6.4. Intermodal Dispersion	312

APPENDIX 1		**319**
APPENDIX 2		**321**
APPENDIX 3		**323**

INDEX	**325**

CHAPTER 1

Introduction

1.1 HISTORICAL DEVELOPMENT OF LIGHTWAVE TECHNOLOGY

The study of light is perhaps one of man's oldest scientific endeavors. Ancient philosophers speculated about its nature and were familiar with the concepts of reflection, refraction, as well as the rectilinear propagation of light. The first systematic writings, in fact, appear to be due to Greek philosophers and mathematicians such as Empedocles (c. 490–430 B.C.) and Euclid (c. 300 B.C.). Although these early studies were more philosophical than scientific, by the time of Galileo Galilei (1564–1642), the power of the experimental method in optics had been demonstrated and the field of classical optics was elevated to the level of a science.

There have been, of course, numerous other scientific milestones achieved in the process of developing modern lightwave technology. However, from a historical perspective there is one other particularly noteworthy date, the year 1880, in which Alexander Graham Bell invented the "photophone." The photophone was a device that varied the intensity of sunlight incident upon it in response to the amplitude of speech vibrations. A receiver could then be used to reconvert the light variations into an electrical signal by means of a selenium detector, and subsequently back into sound. While the photophone itself was impractical due to the rapidity with which its optical signal intensity weakened with propagation distance, the concept of optical communication that it demonstrated is in many ways responsible for the development of the field of integrated optics.

Modern interest in lightwave technology for communication purposes dates from the first demonstration of the laser in 1960. This device, which emits an essentially monochromatic beam of radiation in the visible or near-infrared region, opened up a new portion of the electromagnetic spectrum with frequencies 10,000 times higher than those commonly available in radio communication systems. Because information capacity increases directly with frequency, the laser potentially offers a four order of magnitude increase in available bandwidth, pushing usable frequencies upward from 10 gigahertz (10×10^9 Hz) to 100 terahertz (100×10^{12} Hz). By using only a small portion of the available frequency spectrum, a single laser could in principle carry simultaneously every telephone conversation in North America.

In spite of the tremendous potential bandwidth offered by the laser, strong absorption by rain, snow, fog, and smog prevented the propagation of lasers through the atmosphere. It was not until the development of low-loss optical fibers that optical communications could become a reality. Their development, along with that of compact

1

single-mode semiconductor lasers, has led to the demonstration of communication systems that can transmit information at a rate of over two billion bits per second over 130 kilometers (km) with an error rate of one per billion bits. At this information and error rate, the text of five-entire sets of a 30-volume encyclopedia could be transferred between New York and Philadelphia within a second and the only error might be that of two letters in the text capitalized instead of lower cased.

How, then, has the rapid development in the area of optical communications influenced the field of integrated optics? With the capability to transmit optical signals also came the necessity to periodically reamplify and recondition them by the use of "repeaters" spaced every so often along the transmission path. It is here that the classical optics approach was unsatisfactory. The optical telephone repeater typically involved a laser, detector, lenses, and mirrors, spread out on an optical bench. This equipment was sensitive to temperature changes and thermal gradients over the sizable extent of the table, and mechanical vibrations of separately mounted parts. The elegant solution first suggested by S. E. Miller, a researcher at Bell Laboratories, was to miniaturize the repeater, combining all the components onto a single chip and interconnecting them via small optical transmission lines, or waveguides. This was feasible because of another important property of light in the visible and near infrared, its short wavelength. In this portion of the electromagnetic spectrum, the wavelength of light is only of the order of one micron. Therefore, by fabricating the repeater out of components having sizes measured in terms of wavelengths of light, a system with dimensions of centimeters rather than meters was possible.

Although the initial research in integrated optics was primarily directed toward the optical communications area, other potential applications for combining the unique properties of light into an extremely small package were apparent. As the term is presently construed, integrated optics encompasses many topics. These include, but are not limited to, optical waveguiding, switching, modulation, filtering, interferometry, signal processing, waveguide coupling, optical generation, and detection. To support these functions, integrated optics involves a number of technologies brought together. In particular, the development of micron and submicron fabrication technology in the semiconductor area has made a significant impact on the pace of progress in integrated optics. These advances make possible the combining of optical, electro-optical, and electrical components on the same chip, thereby increasing both the flexibility and scope of integrated-optics components.

1.2 INTEGRATED-OPTICS COMPONENTS

To give the reader a "feel" for the subject of integrated optics, this section describes a number of the types of devices that have been or are presently being developed for both communications and signal processing applications. A detailed analysis of the electromagnetic aspects relevant to their operation is the subject of the remaining chapters of this text. Although in no way exhaustive, it is believed that the devices

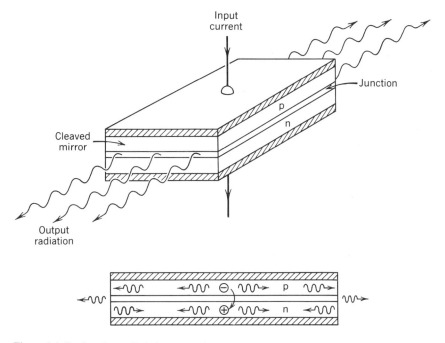

Figure 1.1 Semiconductor diode laser.

chosen for analysis demonstrate many of the important concepts relevant to integrated optics.

1.2.1 Semiconductor Lasers and Detectors

The heart of the optical communications system, as it is for any communications system, is the source. For lightwave applications, the semiconductor laser acts as an optical equivalent to the rf or microwave oscillator. In its simplest form, it consists of a p-n junction diode in which the faces perpendicular to the junction plane are cleaved smooth and parallel, as shown in Figure 1.1. The top and bottom faces, which are parallel to the junction, are contacted with metal electrodes. Upon forward biasing, electrons from the n region are injected across the junction where they combine with holes, giving off radiation at a wavelength determined by the semiconductor's band gap. If the resulting light is directed along an axis perpendicular to the diode mirror faces, it can bounce back and forth, stimulating more electrons to recombine and thereby amplifying the radiation. Because the mirrors have a small transmission coefficient, a portion of the radiation escapes the diode laser and propagates outward as a narrow beam of light.

To sense the emitted radiation, some type of detector is required. One of the simplest practical semiconductor detectors is simply a reversed-biased p-n diode. Incoming radiation incident near the diode junction region and having a sufficiently short wavelength creates hole–electron pairs via the photoelectric effect. These pairs are separated

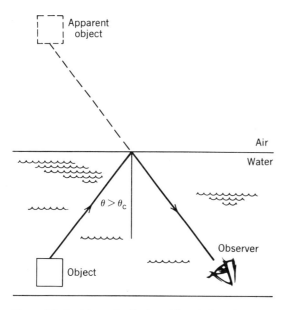

Figure 1.2 Total internal reflection of light incident from a more dense medium upon a less dense one.

and swept across the junction in opposite directions by the built-in junction potential and are subsequently collected as a current by the external circuitry.

1.2.2 Dielectric Waveguides

The principle of optical confinement using a high dielectric material is based upon the phenomenon of total internal reflection (TIR). A common example of TIR is that observed at the bottom of a swimming pool. As shown in Figure 1.2, for observation angles θ greater than the critical angle θ_c the top surface of the pool acts as a mirror, reflecting all light from below back toward the pool bottom. This phenomenon is a direct result of the fact that water has a higher refractive index than air. Suppose next that two such interfaces exist, as is the case for a simple glass microscope slide. Then light generated between the high refractive-index surfaces at an angle $\theta > \theta_c$ will continue to bounce between the upper and lower glass–air interfaces through the process of TIR, as shown in Figure 1.3. This example is a simple version of an optical or

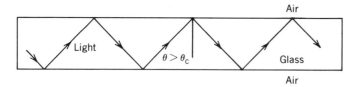

Figure 1.3 A simple dielectric waveguide. Guidance occurs through total internal reflection at upper and lower interfaces.

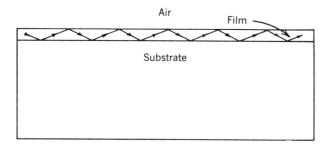

Figure 1.4 Typical geometry for a practical planar slab dielectric waveguide.

dielectric waveguide. More practical geometries for integrated optics can be fabricated, for example, by sputtering a thin, high refractive-index layer onto a lower refractive-index substrate. This yields a guiding film wedged between the lower refractive-index substrate and air regions, as shown in Figure 1.4. When such a guide is made from semiconducting materials, the additional potential exists for integrating electronic and electro-optic components in proximity with the optical ones.

1.2.3   Modulators and Interferometers

For information to be placed upon the laser carrier frequency, the beam must be modulated in some fashion. One particular scheme involving the so-called Mach Zehnder interferometer makes use of the coherent nature of laser light. As shown in Figure 1.5, the input signal is split equally between two waveguiding branches which form the interferometric arms. A substrate material is chosen that is electro-optic, that is,

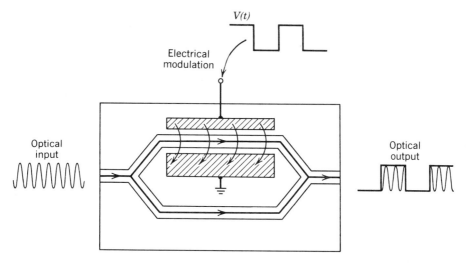

Figure 1.5 Integrated-optics electro-optic modulator. Electric field-induced changes in permittivity result in constructive or destructive interference between signals traveling in the two waveguide arms.

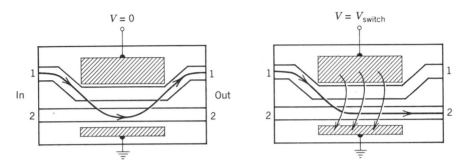

Figure 1.6 Integrated-optics directional coupler. Electric field-induced changes in permittivity cause transfer of a guided signal from one waveguide to the other.

one that allows the optical velocity or equivalently its wavelength to be modified by the application of a dc or low-frequency electric field. One arm of the interferometer has an electric field placed across it whose amplitude may be varied. When no field is applied, the signals from each arm travel the same electrical distance and therefore recombine in phase. When a sufficiently large electric field is applied, the signals recombine 180° out of phase and therefore cancel. Thus, by varying the amplitude of the low-frequency electrical signal, the amplitude of the laser radiation can be modulated. This scheme can therefore be used to place a modulated signal on the laser carrier. It should be noted that the interferometric principle described here is also useful for several other functions including high-speed analog-to-digital conversion, optical phase shifting, and the measurement of such physical parameters as temperature, pressure, and magnetic field.

1.2.4 Directional Couplers

When two identical waveguides are placed in proximity, the power carried by these structures is found to periodically exchange between them with distance, as shown in Figure 1.6. Thus, for example, if the input to one guide is excited with an optical signal while the other is not, then after some distance it is found that all the power has been transferred to the opposite waveguide. The distance required for the exchange depends on the waveguide spacing and the velocity of light along the guides. This velocity-dependent coupling length is the principle behind the operation of the electro-optic switch. With reference to Figure 1.6, let us assume that the waveguides are made from an electro-optic material. With no voltage applied, the coupling length is chosen to be exactly an even number of exchange periods, so that all optical power incident on waveguide 1 exits from the same waveguide. By application of the appropriate dc bias, the velocity within the guides is modified by the electro-optic effect, changing the coupling length by exactly one period and causing the optical radiation to exit from waveguide 2. This technique allows, for example, a fiber optics switching network to be created in which light can be transferred from any incoming optical fiber to any outgoing fiber as shown in Figure 1.7.

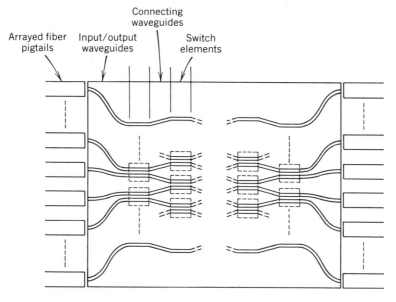

Figure 1.7 Practical implementation of an integrated-optics fiber-switching network. (After Kondo *et al.* (Ref. 5). Copyright © 1982, IEEE. Reproduced by permission.)

1.2.5   Gratings and Filters

If an optical signal is passed through a waveguide having a grating consisting of a series of periodic grooves, the undulations in refractive index cause a small portion of the signal to be reflected at each groove. As shown in Figure 1.8, if the optical wavelength is λ and the groove periodicity is Λ, then the reflected signals will add in phase only when the distance from one groove to the next and then back again is equal to a wavelength. That is,

$$\lambda = 2\Lambda \tag{1.1}$$

For long gratings having small reflectivity per groove, the spread in wavelength $\Delta\lambda$ over which strong reflection occurs can be extremely small, on the order of a few angstroms (1 Å = 1 × 10^{-10} m). Thus, such a grating can be used as either a notch filter in transmission or a narrow band-pass filter in reflection. In the latter case, a series of such filters with slightly different periodicities could be used, for example, to demultipliex a number of optical signals carried on the same fiber, each having a different carrier wavelength. Additionally, such gratings are useful as the end reflectors for semiconductor lasers.

1.2.6   Integrated-Optics Spectrum Analyzer

The spectrum analyzer on a chip is an example of a combination of the technologies described above. With reference to Figure 1.9, the guided beam from a semiconductor

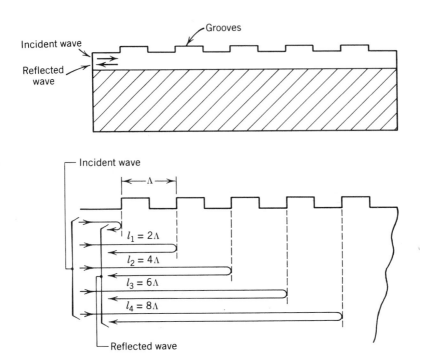

Figure 1.8 Distributed grating reflector. Small reflections from a large number of grooves add up coherently over the length of the grating to strongly reflect the incident wave.

laser is collimated by the use of a planar lens and passed through a grating having periodicity Λ. When the incident optical wave intercepts the grating, it reflects off the grating planes as shown. As is described in Chapter 9, for reflections from all grating planes to add in phase the optical wavelength λ must satisfy the relation

$$\lambda = 2\Lambda \sin \theta \tag{1.2}$$

which also determines the angle θ at which the beam is scattered.

To produce the periodic grating, a portion of the spectrum analyzer chip is fabricated from a piezoelectric material; that is, one for which application of an electric field produces a stress. When an rf signal is applied to a set of interdigital metal electrodes fabricated on the piezoelectric region, the alternating electric field generates a periodic, time-varying stress and an acoustic wave is launched along the surface. The alternating compressions and rarefractions of this surface acoustic wave, or SAW, produce a small, periodic modification to the refractive index of the waveguiding region. The wavelength Λ of the disturbance is inversely proportional to the applied radio frequency.

Since Λ can be changed by varying the radio frequency applied to the SAW transducer, different radio frequencies result in different scattering angles for the guided optical beam. Further, the amplitude of the scattered beam is proportional to the amplitude of the rf signal. If a second lens is used to collimate the scattered light and

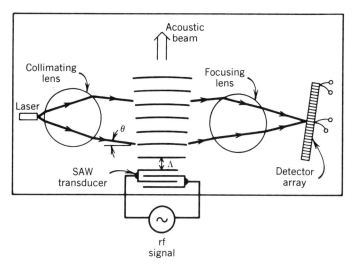

Figure 1.9 Integrated-optics spectrum analyzer showing semiconductor laser source, planar lenses, acoustic grating, and integrated photodiode array.

a series of optical detectors are placed in a line at the output focal plane, there will be a one-to-one correspondence between the amplitude and frequency of the applied rf signal and the generated voltage amplitude and position of the particular detector receiving the optical signal. Thus, if the rf signal contains a number of Fourier components of varying amplitude and frequency, the spectrum will be mapped at the detector array by a corresponding variation in voltage amplitude versus element linear position as shown in Figure 1.10. The chip therefore acts as a discrete Fourier spectrum analyzer.

1.3 ORGANIZATION OF THIS BOOK

Chapter 1 provides an introduction and motivational survey of the area of lightwave technology and integrated optics. Chapter 2 introduces Maxwell's equations, the concepts of power flow, time-harmonic fields, plane-wave propagation, polarization, and group and phase velocity. The topic of duality, which is of great utility in obtaining field solutions for dielectric waveguides, is also presented.

Chapter 3 deals with TE and TM wave incidence on a dielectric half space, with emphasis on those principles most important to optical guidance, including total internal reflection and the Goos–Haenchen shift.

Chapter 4 treats in great detail the slab dielectric waveguide. TE waves on the symmetric slab are analyzed thoroughly and the concepts of modes, the guidance condition, dispersion and cutoff, both from ray and field interpretations, are developed in a physical manner. Once a good physical understanding has been developed, the methodology is extended to the more complicated cases of TE and TM modes on the asymmetric slab. Radiation modes are dealt with in a unique fashion in this text. They

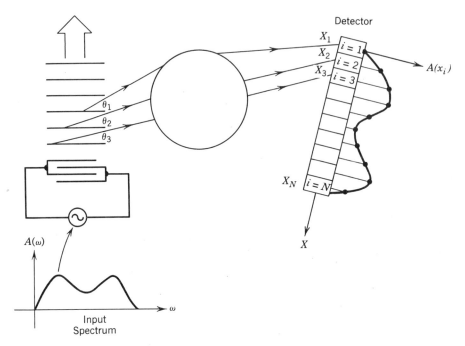

Figure 1.10 Correspondence between a rf signal Fourier spectrum and the position and intensity of diffracted light.

are treated as a logical extension to the guided-wave solutions, obtained by allowing the spacing between perfectly conducting plates located above and below the guiding dielectric layer to approach infinity. It is believed that this technique provides both conceptual advantages in relating radiation modes to guided modes as well as mathematical simplification with regard to mode orthogonality and mode expansion.

Chapter 5 extends the principles developed in analyzing the slab dielectric waveguide to more practical waveguiding geometries. The WKB method is introduced in an intuitive fashion with a minimum of mathematical derivation and is used to calculate propagation constants and spatial extent of modes on graded-index waveguides. The effective index method is presented as a means to analyze a number of transversally confined or so-called three-dimensional waveguides including the strip, rib, and strip-loaded structures. Additionally, the radiation bending losses that invariably accompany curved sections of 3-D waveguides are treated.

Chapter 6 studies the prism coupler in detail, not only as a device for transferring energy into and out of dielectric waveguides but also as a tool for measuring waveguide film properties. Perhaps even more importantly, the prism coupler is used as a vehicle by which the concepts of frustrated total internal reflection, ray superposition, synchronism, and mode selectivity are developed.

In Chapter 7, a number of the methods used for fabricating integrated-optics dielectric and semiconductor waveguides are introduced. Emphasis is placed upon general processing techniques that are not likely to become rapidly obsolete. These include sputtering, diffusion, ion implantation, liquid-phase epitaxy, chemical-vapor deposition, molecular-beam epitaxy, ion and plasma etching, as well as flame hydrolysis for optical fibers.

Chapter 8 derives the Lorentz reciprocity theorem and uses the resulting expression in a unified manner to treat a number of important coupled-mode problems. The concept of forward coupling between waveguides is analyzed in detail using the electro-optic switch as an example. Backward coupling is studied in the context of a slab waveguide perturbed by a periodic grating.

Chapter 9 investigates the phenomenon of Bragg scattering in detail. Physical insight is first developed by consideration of deflection of an optical plane wave by a bulk acoustic wave. Use is made of the ray superposition method developed in an earlier chapter. The analysis is further extended to consider deflection by finite aperture bulk optical and acoustic beams and used to treat the bulk-wave acousto-optic spectrum analyzer. Deflection efficiency is determined quantitatively by presenting the concepts of strain and the acousto-optic effect. Finally, the Lorentz reciprocity theorem developed in Chapter 8 is readily extended to treat the guided-wave spectrum analyzer.

Chapter 10 provides an introductory analysis of wave propagation along optical fibers. It represents a significant departure from standard descriptions used in most texts. Simple approximate solutions for single-mode fibers with several different refractive-index profiles are stressed rather than exact or approximate solutions to Maxwell's equations for the step-index profile. The method avoids the solution of complicated transcendental equations and the introduction of Bessel functions. It further provides simple analytical solutions for field and power density, group velocity, and waveguide pulse distortion. The emphasis on the single- rather than multimode fiber seems particularly justified since the former is the most likely candidate for long-distance high-capacity communication links. Additionally, for completeness, an analysis of linearly polarized waves for multimode guides is included as an advanced section. Chapter 10 also extends the concept of dispersion introduced in an elementary fashion in Chapter 2 to include the effect of pulse distortion. Material, waveguide, and intermodal dispersion are all treated within the context of optical-fiber pulse transmission.

REFERENCES

1. Born, M., and Wolf, E. *Principles of Optics.* 4th ed. New York: Pergamon, 1970.
2. Conwell, E. M. "Integrated optics." *Physics Today* (May 1976): 48–59.
3. Glenn, A. B., and Keiser, G. E. "A perspective of lightwave technology." *IEEE Communications* 23 (1985): 8–9.
4. Hecht, J. "Victorian experiments and optical communications." *IEEE Spectrum* (Feb. 1985): 69–73.
5. Kondo, M., *et al.* "Integrated optical switch matrix for single-mode fiber networks." *IEEE Transactions on Microwave Theory Technology* MTT-30 (1982): 1747–1753.

CHAPTER 2

Basic Electromagnetic Theory

2.1 MAXWELL'S EQUATIONS

To analyze wave propagation in an isotropic medium we begin with a statement of Maxwell's equations:

$$\nabla \times \mathbf{E}(\mathbf{r}, t) = -\frac{\partial}{\partial t} \mathbf{B}(\mathbf{r}, t) \tag{2.1}$$

$$\nabla \times \mathbf{H}(\mathbf{r}, t) = \frac{\partial}{\partial t} \mathbf{D}(\mathbf{r}, t) + \mathbf{J}(\mathbf{r}, t) \tag{2.2}$$

$$\nabla \cdot \mathbf{D}(\mathbf{r}, t) = \rho(\mathbf{r}, t) \tag{2.3}$$

$$\nabla \cdot \mathbf{B}(\mathbf{r}, t) = 0 \tag{2.4}$$

In the above relations, the field quantities **E** and **H** represent, respectively, the electric and magnetic fields, and **D** and **B** the electric and magnetic displacements. **J** and ρ represent the current and charge sources. Further, the position vector **r** defines a particular location in space (x, y, z) at which the field is being measured. That is,

$$\mathbf{E}(x, y, z, t) \equiv \mathbf{E}(\mathbf{r}, t)$$

An auxiliary relation, called the continuity equation, relates the current and charge sources **J** and ρ.

$$\nabla \cdot \mathbf{J}(\mathbf{r}, t) + \frac{\partial}{\partial t} \rho(\mathbf{r}, t) = 0 \tag{2.5}$$

It is important to note that Eqs. 2.1–2.5 are not all independent. In fact, only Eqs. 2.1, 2.2, and 2.5 are needed to derive Eqs. 2.3 and 2.4. This is readily seen by taking the divergence of Eqs. 2.1 and 2.2. Taking $\nabla \cdot$ (Eq. 2.1) yields

$$\nabla \cdot (\nabla \times \mathbf{E}) = -\frac{\partial}{\partial t} (\nabla \cdot \mathbf{B})$$

where we have freely interchanged the spatial and temporal operators, $\nabla \cdot$ and $\partial/\partial t$. Note, however, that the divergence of the curl of any vector is identically zero. This implies from above that the divergence of $\mathbf{B}$ is at most equal to a constant which is independent of time. If such a "background" field exists, it has been with us since creation and is of little interest to any kind of time-varying problem. Therefore, we arbitrarily set the constant equal to zero, thus yielding Eq. 2.4. If we take the divergence of Eq. 2.2 we obtain

$$\nabla \cdot (\nabla \times \mathbf{H}) = \frac{\partial}{\partial t} (\nabla \cdot \mathbf{D}) + \nabla \cdot \mathbf{J}$$

However, the continuity equation 2.5 shows that the divergence of $\mathbf{J}$ is equal to minus the time derivative of the charge density ρ. Thus, to within an arbitrary constant, we obtain Eq. 2.3.

2.2 CONSTITUTIVE RELATIONS FOR AN ISOTROPIC MEDIUM

An examination of Maxwell's curl equations, 2.1 and 2.2, indicates that there are a total of 6 equations (one for each vector component) and 12 unknowns (one for each component of $\mathbf{E}$, $\mathbf{H}$, $\mathbf{B}$, and $\mathbf{D}$). Note that $\mathbf{J}$ represents a source term which is considered to be given. It is the constitutive relations that provide the additional constraints needed to solve Eqs. 2.1 and 2.2. These relations characterize a given material on a macroscopic level and are described in terms of the two scalar quantities, the permeability μ and the permittivity ϵ:

$$\mathbf{B} = \mu\mathbf{H} \tag{2.6}$$

$$\mathbf{D} = \epsilon\mathbf{E} \tag{2.7}$$

In free space, $\mu = \mu_0 = 4\pi \times 10^{-7}$ henrys/meter (H/m), $\epsilon = \epsilon_0 \cong 8.85 \times 10^{-12}$ farads/meter (F/m). Note that Eqs. 2.6 and 2.7 represent a set of six equations, one for each field component.

2.3 TIME-HARMONIC SOLUTION OF MAXWELL'S EQUATIONS

The form of Maxwell's equations given by Eqs. 2.1–2.4 is rather cumbersome because of the inclusion of both time and space dependence. A great deal of simplification is possible if we are dealing with time-harmonic fields, that is, fields varying at a sinusoidal frequency ω. If we designate such a field as $\mathbf{A}(\mathbf{r}, t)$ then, for example, its $\hat{x}$-

component can be represented as

$$A_x(\mathbf{r}, t) = A_x(\mathbf{r}) \cos[\omega t + \phi_x(\mathbf{r})]$$

$$= \text{Re}[A_x(\mathbf{r}) \, e^{j\phi_x(\mathbf{r})} e^{j\omega t}]$$

$$= \text{Re}[\underline{A}_x(\mathbf{r}) e^{j\omega t}]$$

where the complex phasor quantity $\underline{A}_x(\mathbf{r})$ is defined by

$$\underline{A}_x(\mathbf{r}) = A_x(\mathbf{r}) e^{j\phi_x(\mathbf{r})}$$

The $\hat{y}$ and $\hat{z}$ components of the field are similarly described and the total field may, therefore, be written as

$$\mathbf{A}(\mathbf{r}, t) = \text{Re}[\underline{\mathbf{A}}(\mathbf{r}) e^{j\omega t}] \tag{2.8}$$

where

$$\underline{\mathbf{A}}(\mathbf{r}) = \hat{x} \, \underline{A}_x(\mathbf{r}) + \hat{y} \, \underline{A}_y(\mathbf{r}) + \hat{z} \, \underline{A}_z(\mathbf{r})$$

To obtain Maxwell's equations for time-harmonic fields, we assume all field quantities to be of the form given by Eq. 2.8. That is,

$$\begin{Bmatrix} \mathbf{E}(\mathbf{r}, t) \\ \mathbf{H}(\mathbf{r}, t) \\ \mathbf{D}(\mathbf{r}, t) \\ \mathbf{B}(\mathbf{r}, t) \end{Bmatrix} = \text{Re} \left\{ \begin{bmatrix} \underline{\mathbf{E}}(\mathbf{r}) \\ \underline{\mathbf{H}}(\mathbf{r}) \\ \underline{\mathbf{D}}(\mathbf{r}) \\ \underline{\mathbf{B}}(\mathbf{r}) \end{bmatrix} e^{j\omega t} \right\}$$

The appropriate field quantities are then substituted into Maxwell's equations 2.1–2.4. Thus, for example, Eq. 2.1 becomes

$$\nabla \times \text{Re}[\underline{\mathbf{E}}(\mathbf{r}) e^{j\omega t}] = -\partial/\partial t \, \text{Re}[\underline{\mathbf{B}}(\mathbf{r}) e^{j\omega t}] \tag{2.9}$$

To simplify this expression, we need to show that it is possible to move the operators $\nabla \times$ and $\partial/\partial t$ inside the indicated square brackets. That is, we wish to show that Eq. 2.9 is equivalent to

$$\text{Re}[\nabla \times \underline{\mathbf{E}}(\mathbf{r}) e^{j\omega t}] = -\text{Re}[\partial/\partial t \, \underline{\mathbf{B}}(\mathbf{r}) e^{j\omega t}] \tag{2.10}$$

Let us first demonstrate the validity of the interchange of $\partial/\partial t$ and Re operators. Now

$$\partial/\partial t \, \text{Re}[\underline{\mathbf{B}} e^{j\omega t}]$$

can be written using the complex conjugate (*) as

$$\partial/\partial t \; \tfrac{1}{2}[\underline{\mathbf{B}}e^{j\omega t} + (\underline{\mathbf{B}}e^{j\omega t})^*]$$

$$= \frac{1}{2}\left[\left(\frac{\partial}{\partial t}\underline{\mathbf{B}}e^{j\omega t}\right) + \left(\frac{\partial}{\partial t}\underline{\mathbf{B}}e^{j\omega t}\right)^*\right]$$

$$= \mathrm{Re}\left(\frac{\partial}{\partial t}\underline{\mathbf{B}}e^{j\omega t}\right)$$

which is the desired result. Note that we have used the fact that the variable t is real so that $\partial/\partial t^* = \partial/\partial t$.

To demonstrate the validity of interchange of the $\nabla \times$ and Re operators, it is sufficient to note that ∇ can be represented as the sum of three partial spatial derivatives, each of which is a real variable. That is,

$$\nabla = \hat{x}\,\partial/\partial x + \hat{y}\,\partial/\partial y + \hat{z}\,\partial/\partial z = \nabla^*$$

Thus

$$\nabla \times \mathrm{Re}[\underline{\mathbf{E}}e^{j\omega t}]$$

$$= \nabla \times \tfrac{1}{2}[\underline{\mathbf{E}}e^{j\omega t} + (\underline{\mathbf{E}}e^{j\omega t})^*]$$

$$= \tfrac{1}{2}[\nabla \times \underline{\mathbf{E}}e^{j\omega t} + (\nabla \times \underline{\mathbf{E}}e^{j\omega t})^*]$$

$$= \mathrm{Re}[\nabla \times \underline{\mathbf{E}}e^{j\omega t}]$$

We have therefore demonstrated the validity of Eq. 2.10, which can be written in the form

$$\mathrm{Re}\{[\nabla \times \underline{\mathbf{E}}(\mathbf{r}) + j\omega\underline{\mathbf{B}}(\mathbf{r})]e^{j\omega t}\} = 0 \tag{2.11}$$

Let us show that the above expression requires the quantities inside the square brackets to be equal to zero. We observe that these quantities can be represented as a complex phasor, $\underline{\mathbf{A}} = \mathbf{A}' + j\mathbf{A}''$, where $\mathbf{A}'$ and $\mathbf{A}''$ represent the real and imaginary parts of $\underline{\mathbf{A}}$, respectively. Thus Eq. 2.11 is of the form

$$\mathrm{Re}\{[\mathbf{A}' + j\mathbf{A}'']e^{j\omega t}\} = 0 \tag{2.12}$$

Consider now, the solution to the expression above at the two instants of time defined by $\omega t = 0$ and $\omega t = \pi/2$. At $\omega t = 0$ we require that $\mathbf{A}' = 0$ whereas at $\omega t = \pi/2$, $\mathbf{A}''$ must be zero. For Eq. 2.12 to be valid for *all times* we must therefore demand that $\mathbf{A}' = \mathbf{A}'' = 0$ and thus $\underline{\mathbf{A}} = 0$. Thus, we are led to the conclusion that the first of Maxwell's equations can be written compactly as

$$\nabla \times \mathbf{E}(\mathbf{r}) = -j\omega \underline{\mathbf{B}}(\mathbf{r})$$

The effect of the time-harmonic assumption has been to replace the true time–space fields with their phasor amplitudes and to replace the partial derivative $\partial/\partial t$ with $j\omega$. Following this prescription for the remaining equations 2.2–2.4 therefore yields for the time-harmonic form of Maxwell's equations:

$$\nabla \times \underline{\mathbf{E}}(\mathbf{r}) = -j\omega \; \underline{\mathbf{B}}(\mathbf{r}) \tag{2.13}$$

$$\nabla \times \underline{\mathbf{H}}(\mathbf{r}) = j\omega \; \underline{\mathbf{D}}(\mathbf{r}) + \underline{\mathbf{J}}(\mathbf{r}) \tag{2.14}$$

$$\nabla \cdot \underline{\mathbf{D}}(\mathbf{r}) = \underline{\rho}(\mathbf{r}) \tag{2.15}$$

$$\nabla \cdot \underline{\mathbf{B}}(\mathbf{r}) = 0 \tag{2.16}$$

Note that the effect of the time-harmonic assumption has been to eliminate the time dependence from Maxwell's equations. While it may appear that this simplification has been obtained at the sacrifice of requiring analysis of time-harmonic fields only, recall that any arbitrary time-varying quantity can be analyzed in terms of a super-position of Fourier components. Thus, if the field quantity $\mathbf{A}(\mathbf{r}, t)$ is not sinusoidal we can represent it as

$$\mathbf{A}(\mathbf{r}, t) = \text{Re}\left[\int_{-\infty}^{\infty} d\omega \; \underline{\mathbf{A}}(\mathbf{r}, \omega)e^{j\omega t}\right] \tag{2.17}$$

The solutions to Maxwell's equations can be obtained for each Fourier component, $\underline{\mathbf{A}}(\mathbf{r}, \omega)e^{j\omega t}$, and the final result summed again over ω.

Example 2.1

Represent the following electric field in phasor form:

$$\underline{\mathbf{E}}(\mathbf{r}, t) = \hat{x} \cos(\omega t - kz) + 2\hat{y} \sin(\omega t - kz)$$

Solution

The electric field can be written as

$$\underline{\mathbf{E}}(\mathbf{r}, t) = \hat{x} \cos(\omega t - kz) + 2\hat{y} \cos(\omega t - kz - \pi/2)$$

$$= \text{Re}[(\hat{x}e^{-jkz} + 2\hat{y}e^{-j\pi/2}e^{-jkz})e^{j\omega t}]$$

Thus we identify

$$\underline{E}_x(\mathbf{r}) = E_x e^{j\phi_x}$$

where

$$E_x = 1 \qquad \phi_x = -kz$$

and

$$\underline{E}_y(\mathbf{r}) = E_y e^{j\phi_y}$$

where

$$E_y = 2 \qquad \phi_y = -(kz + \pi/2)$$

The complete phasor $\underline{\mathbf{E}}(\mathbf{r})$ can therefore be written as

$$\underline{\mathbf{E}}(\mathbf{r}) = (\hat{x} - j2\hat{y})e^{-jkz}$$

2.4   THE WAVE EQUATION IN SOURCE-FREE ISOTROPIC MEDIA

We wish to determine the types of solutions to Maxwell's equations which are allowed in a source-free isotropic region. Note that by ''source free'' we do not require that $\underline{\mathbf{J}}(\mathbf{r})$ and $\underline{\rho}(\mathbf{r})$ are identically zero but only that we are in a region over which $\mathbf{J}$ and ρ do not extend, as shown in Figure 2.1.

2.4.1   The Wave Equation

To obtain a description of the type of electromagnetic waves that may exist in a source-free medium we make use of the two curl equations 2.13 and 2.14 with the source term $\underline{\mathbf{J}}(\mathbf{r})$ set equal to zero. We additionally introduce the constitutive relations 2.6 and 2.7 to eliminate the variables $\underline{\mathbf{B}}(\mathbf{r})$ and $\underline{\mathbf{D}}(\mathbf{r})$ yielding

$$\nabla \times \underline{\mathbf{E}}(\mathbf{r}) = -j\omega\mu\underline{\mathbf{H}}(\mathbf{r})$$

$$\nabla \times \underline{\mathbf{H}}(\mathbf{r}) = j\omega\epsilon\underline{\mathbf{E}}(\mathbf{r})$$

By taking the curl of the first equation above and substituting into the second expression we can eliminate $\underline{\mathbf{H}}(\mathbf{r})$ yielding

$$\nabla \times [\nabla \times \underline{\mathbf{E}}(\mathbf{r})] = -j\omega\mu \, \nabla \times \underline{\mathbf{H}}(\mathbf{r}) = \omega^2\mu\epsilon \, \underline{\mathbf{E}}(\mathbf{r})$$

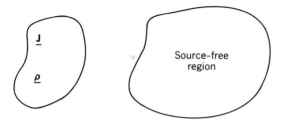

Figure 2.1 Location of sources and region of analysis.

Using the vector identity,

$$\nabla \times \nabla \times \underline{E}(\mathbf{r}) = \nabla [\nabla \cdot \underline{E}(\mathbf{r})] - \nabla^2 \underline{E}(\mathbf{r})$$

and noting that since $\rho = 0$ and ϵ is a scalar constant,

$$\nabla \cdot \underline{D}(\mathbf{r}) = \nabla \cdot \epsilon \underline{E}(\mathbf{r}) = \epsilon \nabla \cdot \underline{E}(\mathbf{r}) = 0$$

then we obtain

$$\nabla^2 \underline{E}(\mathbf{r}) + \omega^2 \mu \epsilon \underline{E}(\mathbf{r}) = 0 \tag{2.18}$$

Relation 2.18 is known as the wave equation. It represents three equations of identical form, one for each component of $\underline{E}(\mathbf{r})$. For example, the $\hat{x}$ component of $\underline{E}$ satisfies the relation

$$\left(\frac{\partial^2}{\partial x^2} + \frac{\partial^2}{\partial y^2} + \frac{\partial^2}{\partial z^2} \right) \underline{E}_x(\mathbf{r}) + \omega^2 \mu \epsilon \underline{E}_x(\mathbf{r}) = 0$$

2.4.2 The Dispersion Relation

Let us attempt to find a solution for $\underline{E}_x(\mathbf{r})$ above of the form

$$\underline{E}_x(\mathbf{r}) = \underline{E}_{x0} \exp[-j (k_x x + k_y y + k_z z)] \tag{2.19}$$

where $\underline{E}_{x0}$ is an arbitrary constant. We note that due to the exponential form chosen for $\underline{E}_x(\mathbf{r})$, the effect of operating $\partial^2/\partial x^2$ on $\underline{E}_x(\mathbf{r})$ is to yield $-k_x^2 \underline{E}_x(\mathbf{r})$ with analogous results for partials with respect to y and z. Therefore, substitution of the assumed form of $\underline{E}_x$ into our differential equation yields

$$k_x^2 + k_y^2 + k_z^2 = \omega^2 \mu \epsilon \equiv k^2 \tag{2.20}$$

Equation 2.20 is known as the dispersion relation. Since the $\hat{y}$ and $\hat{z}$ components for $\underline{E}(\mathbf{r})$ are solutions of the same wave equation they must also have the same dispersion relation. Note additionally that the amplitude of the wave, $\underline{E}_{x0}$, is arbitrary.

2.4.3   The Wave Vector k

The solution to our wave equation $\underline{E}_x(\mathbf{r})$ can be given physical significance if we define the wave vector **k** by

$$\mathbf{k} \equiv \hat{x}k_x + \hat{y}k_y + \hat{z}k_z$$

Noting that the vector **r** from the origin of our coordinate system to any observation point (x, y, z) is given by

$$\mathbf{r} = \hat{x}x + \hat{y}y + \hat{z}z$$

then $\underline{E}_x(\mathbf{r})$ can be represented compactly as

$$\underline{E}_x(\mathbf{r}) = \underline{E}_{x0}e^{-j\mathbf{k}\cdot\mathbf{r}}$$

and the true time-varying x component of the electric field is given by

$$E_x(\mathbf{r}, t) = \mathrm{Re}\left\{\underline{E}_{x0} \exp\left[j\omega\left(t - \frac{k}{\omega}\hat{k} \cdot \mathbf{r}\right)\right]\right\}$$

To interpret the meaning of the wave vector **k** we take a snapshot of the electric field $E_x(\mathbf{r}, t)$ at some time $t = t_0$ and ask at what points **r** is the amplitude of E_x a constant. Since t is fixed, this requires additionally that $\mathbf{k} \cdot \mathbf{r}$ also be fixed or

$$\hat{k} \cdot \mathbf{r} = \text{constant} \equiv C_1.$$

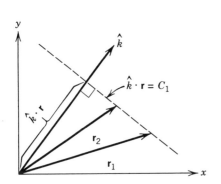

Figure 2.2 Constant phase plane (dashed line).

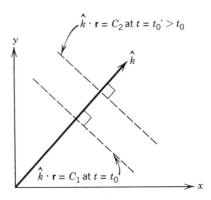

Figure 2.3 Motion of the constant phase plane with time.

This relation is shown diagramatically in Figure 2.2 for the two-dimensional case ($k_z = 0$). We see that requiring $\hat{k} \cdot \mathbf{r}$ to be a constant is equivalent to requiring that the projection of $\mathbf{r}$ onto $\hat{k}$ be a constant. In two dimensions this defines a line perpendicular to $\hat{k}$ and in three dimensions a plane. If we increment t to some new time $t_0' > t_0$ and look for the points in space having the same amplitude as at t_0 we must be at some new plane, as shown in Figure 2.3. The diagram leads to the conclusion that the $\mathbf{k}$ vector represents the propagation direction of the electromagnetic radiation. Thus by specifying k_x, k_y, and k_z we determine in which direction our wave is traveling. Care must be taken in choosing $\mathbf{k}$, however. From Eq. 2.20 we note that the magnitude of $\mathbf{k}$ must satisfy the relation

$$|\mathbf{k}| \equiv k = \omega\sqrt{\mu\epsilon}$$

That is, we are free to choose any propagation direction lying on a sphere of radius k, as shown in Figure 2.4. Radiation of the form described above and given mathematically by Eq. 2.19 is called a plane electromagnetic wave or simply a "plane wave."

Example 2.2

A plane wave oscillating at a frequency $f = 3 \times 10^{14}$ Hz and propagating in free space has an electric field given by

$$\underline{\mathbf{E}}(\mathbf{r}) = \hat{y}e^{-jk_x x}e^{-jk_z z}$$

where

$$k_x = \pi \times 10^6 \text{ m}^{-1} \qquad k_y = 0$$

What is the angle θ that the propagation vector $\mathbf{k}$ makes with respect to the z axis?

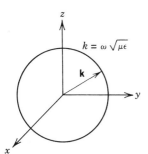

Figure 2.4 The k surface.

Solution

The magnitude of **k** is

$$k = \omega\sqrt{\mu_0\epsilon_0} = 2\pi \times 3 \times 10^{14} \times \sqrt{4\pi \times 10^{-7} \times 8.85 \times 10^{-12}}$$

$$= \frac{2\pi \times 3 \times 10^{14}}{3 \times 10^8} = 2\pi \times 10^6 \text{ m}^{-1}$$

The *z* component of **k** is found from dispersion relation 2.20 with $k_y = 0$.

$$k_z = \sqrt{k^2 - k_x^2}$$

$$= \sqrt{(2\pi)^2 - \pi^2} \times 10^6 \text{ m}^{-1}$$

$$= \sqrt{3}\pi \times 10^6 \text{ m}^{-1}$$

The propagation angle with respect to the *z* axis is therefore obtained from the geometric relation

$$\tan\theta = \frac{k_x}{k_z} = \frac{1}{\sqrt{3}}$$

or

$$\theta = 18.4°$$

2.4.4 Maxwell's Equations for Plane-Wave Solutions

We have shown that the assumption of time-harmonic fields reduced the complexity of Maxwell's equations by eliminating time as a variable. For many problems of interest, the spatial dependence may also be eliminated provided that the field solutions are known a priori to be plane wave in nature. We, therefore, assume time-harmonic fields having a spatial variation of the form given in Eq. 2.19

$$\left.\begin{array}{c}\underline{\mathbf{E}}(\mathbf{r})\\ \underline{\mathbf{H}}(\mathbf{r})\end{array}\right\} = \left.\begin{array}{c}\underline{\mathbf{E}}\\ \underline{\mathbf{H}}\end{array}\right\} e^{-j\mathbf{k}\cdot\mathbf{r}}$$

Let us investigate the effect of the curl operator on such fields.

$$\nabla \times \underline{\mathbf{E}}(\mathbf{r}) = (\hat{x}\partial/\partial x + \hat{y}\partial/\partial y + \hat{z}\partial/\partial z) \times \underline{\mathbf{E}}e^{-j(k_x x + k_y y + k_z z)}$$

$$= -j(\hat{x}k_x + \hat{y}k_y + \hat{z}k_z) \times \underline{\mathbf{E}}e^{-j(k_x x + k_y y + k_z z)}$$

$$= (-j\mathbf{k} \times \underline{\mathbf{E}})e^{-j\mathbf{k}\cdot\mathbf{r}} = -j\mathbf{k} \times \underline{\mathbf{E}}(\mathbf{r})$$

Using the same technique one can easily show that

$$\nabla \cdot \underline{E}(\mathbf{r}) = (-j\,\mathbf{k} \cdot \underline{E})e^{-j\mathbf{k}\cdot\mathbf{r}}$$

Thus, for plane-wave solutions in a source-free region, the ∇ operator is replaced by $-j\mathbf{k}$ yielding a simplified form of Maxwell's equations

$$\mathbf{k} \times \underline{E} = \omega\underline{B} \qquad (2.21)$$

$$\mathbf{k} \times \underline{H} = -\omega\underline{D} \qquad (2.22)$$

$$\mathbf{k} \cdot \underline{D} = 0 \qquad (2.23)$$

$$\mathbf{k} \cdot \underline{B} = 0 \qquad (2.24)$$

Note that all spatial variation has been eliminated due to the fact that the exponential term $e^{-j\mathbf{k}\cdot\mathbf{r}}$ is common to all field quantities. Several general statements can be made by examination of these relations. First, note that expressions 2.23 and 2.24 indicate that both $\underline{D}$ and $\underline{B}$ must be in a plane which is perpendicular to the direction of propagation. Further, for isotropic materials in which $\underline{D}$ and $\underline{B}$ are related to $\underline{E}$ and $\underline{H}$, respectively, by scalars, then relations 2.21 and 2.22 simplify to

$$\mathbf{k} \times \underline{E} = \omega\mu\underline{H}$$

$$\mathbf{k} \times \underline{H} = -\omega\epsilon\underline{E}$$

which implies that $\underline{E}$ and $\underline{H}$ are perpendicular to $\mathbf{k}$ as well as to each other with $\underline{E} \times \underline{H}$ being in the direction of $\mathbf{k}$ as shown in Figure 2.5. If we assume that $\underline{E}$ is polarized along the $\hat{x}$ direction with real amplitude E_{x0} and that $\mathbf{k}$ is along $\mathbf{z}$, then from above

$$\underline{H} = \frac{1}{\omega\mu}(\mathbf{k} \times \underline{E}) = \frac{1}{\omega\mu}(\hat{z}k_z) \times (\hat{x}E_{x0}e^{-jk_z z})$$

$$= \hat{y}\frac{1}{\eta}E_{x0}e^{-jk_z z}$$

where the characteristic impedance η is defined by

$$\eta \equiv \sqrt{\mu/\epsilon} \qquad (2.25)$$

Thus, for a plane wave

$$\frac{|\mathbf{E}|}{|\mathbf{H}|} = \eta \cong 377\ \Omega \text{ in free space}$$

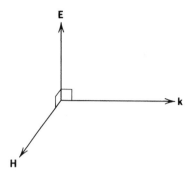

Figure 2.5 Relative orientation of $\underline{E}$, $\underline{H}$, and k vectors.

The true time–space variation for the fields $\mathbf{E}$ and $\mathbf{H}$ may be obtained by multiplying the corresponding complex fields by $e^{j\omega t}$ and taking the real part. Thus,

$$\mathbf{E}(\mathbf{r}, t) = \mathrm{Re}(\hat{x}E_{x0}e^{j\omega t}e^{-jk_z z})$$

$$= \hat{x}E_{x0}\cos(\omega t - k_z z)$$

$$\mathbf{H}(\mathbf{r}, t) = \mathrm{Re}\left(\hat{y}\frac{E_{x0}}{\eta}e^{j\omega t}e^{-jk_z z}\right)$$

$$= \hat{y}\frac{E_{x0}}{\eta}\cos(\omega t - k_z z)$$

Note that $\mathbf{E}$ and $\mathbf{H}$ are everywhere in phase. A "snapshot" of the fields at some arbitrary time t would yield a variation in amplitude with z as shown in Figure 2.6.

2.5 PHASE VELOCITY

We have shown that constant phase planes lie perpendicular to the propagation vector $\mathbf{k}$ and that as time is incremented these planes travel along $\mathbf{k}$. We now determine the velocity at which the constant phase planes propagate.

Without loss of generality, our coordinate system can be oriented so that the $\hat{z}$ axis is along $\mathbf{k}$. Then $k_x = k_y = 0$ and $k_z = k = \omega\sqrt{\mu\epsilon}$. The electric field is then given by

$$\mathbf{E}(z, t) = \mathbf{E}_0\cos(\omega t - k_z z + \phi)$$

which at two particular times t_1 and t_2 looks as shown in Figure 2.7. We note that an arbitrary phase point has traveled or propagated to the right over this time interval. To

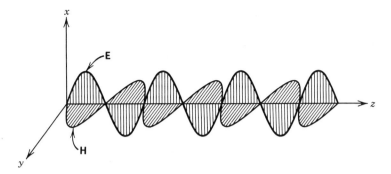

Figure 2.6 Variation of the electric and magnetic field amplitudes with position.

determine the velocity of this arbitrary phase point, we consider an observer moving along at the same velocity as this point. For the observer to remain at the same position with respect to the field requires that the argument of $\mathbf{E}(z, t)$ remains constant. As time progresses, we must therefore stay at a point z defined by

$$\omega t - k_z z + \phi = \text{constant}$$

Taking the derivative with respect to z above determines how fast we must travel along z to remain at the same point of the wave yielding

$$\frac{dz}{dt} = \omega/k_z \equiv v_p \tag{2.26}$$

where v_p is defined as the phase velocity. For free space we have

$$\omega/k_z = 1/\sqrt{\mu_0\epsilon_0} \cong 3 \times 10^8 \text{ m/s}$$

which is the velocity of light in free space.

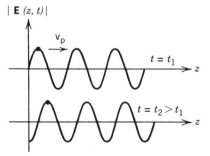

Figure 2.7 Motion of a constant phase point with time.

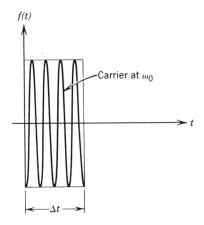

f(t)

Carrier at ω_0

t

Δt

Figure 2.8 Modulated carrier for $z = 0$.

2.6 GROUP VELOCITY

A purely sinusoidal wave such as the one just discussed has neither a beginning nor an end and can, therefore, convey no information. Suppose it is desired to transmit a message, for example, "hello," through some medium. We are, therefore, required to modulate the signal in some fashion. Let the following simple coding be chosen:

A pulse of duration Δt means "hello"

A pulse of duration $2\Delta t$ means "goodbye"

Thus, the message "hello" would look as shown in Figure 2.8. Here $f(t)$ is used to represent the amplitude of the electric or magnetic field at the signal source which is assumed to be located at $z = 0$.

The message is observed to consist of two portions: a relatively slowly varying envelope, in our case a rectangular pulse of duration Δt, and a rapidly varying carrier at radian frequency ω_0. For optical communications such a signal could be generated, for example, either by direct modulation of the drive current of a semiconductor laser or by passing a cw laser beam through an optical switch which is pulsed on for the duration Δt by an applied external voltage. It is the velocity of the envelope rather than the carrier that is of primary interest since this dictates the speed at which the information travels between transmitter and receiver. The spectrum of the message is shown in Figure 2.9 and is observed to consist of a sinc function centered at the carrier frequency ω_0. The half width of the first major lobe of the function is known from simple Fourier transform theory to be proportional to the reciprocal of the pulse duration.

Rather than presenting at this point a general analysis of the transmission characteristics of a signal with such a complicated frequency spectrum, let us instead consider a modulated signal with the very simple spectrum, shown in Figure 2.10. The signal consists of two equal amplitude tones at frequencies $\omega_0 \pm \Delta\omega$. While such a spectrum

Signal Fourier
transform

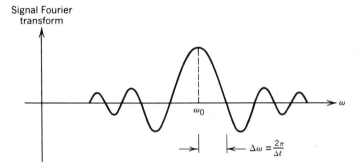

Figure 2.9 Fourier transform of the signal shown in Figure 2.8.

is not a terribly practical one, it introduces the concept of signal velocity in a straight-forward manner. The extension of the analysis presented here for more general signal modulations is treated in detail in Section 6 of Chapter 10.

We therefore represent the modulated signal at the source location by

$$f(t) = \text{Re}[e^{j(\omega_0 + \Delta\omega)t} + e^{j(\omega_0 - \Delta\omega)t}]$$

$$= 2 \cos \omega_0 t \cos \Delta\omega t$$

The signal, shown in Figure 2.11, consists of a carrier at frequency ω_0 modulated by a slowly varying envelope having frequency $\Delta\omega$.

Let us assume that each frequency component of $f(t)$ travels along a propagation direction z with an associated propagation constant $k(\omega)$. By superposition, the received signal $f(t, z)$ at some arbitrary distance z from the source will be

$$f(t, z) = \text{Re}[e^{j(\omega_0 + \Delta\omega)t}e^{-jk(\omega_0 + \Delta\omega)z} + e^{j(\omega_0 - \Delta\omega)t}e^{-jk(\omega_0 - \Delta\omega)z}] \tag{2.27}$$

Assuming that $\Delta\omega$ is sufficiently small, then $k(\omega_0 \pm \Delta\omega)$ can be expanded to good approximation in a first-order Taylor series as

$$k(\omega_0 \pm \Delta\omega) \cong k(\omega_0) \pm \Delta\omega k^{(1)}(\omega_0) \tag{2.28}$$

Simplified signal
fourier transform

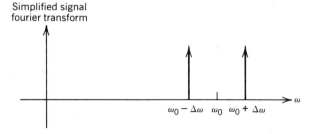

Figure 2.10 Simplified approximation to the Fourier transform of Figure 2.9.

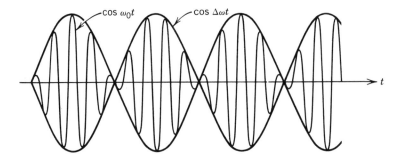

Figure 2.11 Simplified time-domain version of the modulated signal. Figure 2.11 is the inverse Fourier transform of Figure 2.10.

where

$$k^{(1)}(\omega_0) \equiv \frac{dk(\omega)}{d\omega}\bigg|_{\omega=\omega_0}$$

Substituting Eq. 2.28 into Eq. 2.27 above yields

$$f(t, z) = 2\cos[\omega_0 t - k(\omega_0)z]\cos[\Delta\omega(t - k^{(1)}(\omega_0)z)]$$

$$= 2\cos\omega_0(t - \tau_p)\cos\Delta\omega(t - \tau_g)$$

where

$$\tau_p = \frac{k(\omega_0)}{\omega_0} z$$

and

$$\tau_g = k^{(1)}(\omega_0)z$$

The quantities τ_p and τ_g are defined as the phase and group delays, respectively. Note that the received signal is identical with that of the launched one except that the carrier and envelope have been delayed by the times τ_p and τ_g, respectively. We can associate propagation velocities with each of these delays through the relations

$$v_p = z/\tau_p$$

$$v_g = z/\tau_g$$

(2.29)

Thus, in analogy with the monochromatic plane wave, v_p is the phase velocity representing the speed at which the carrier travels from input to output. The envelope velocity v_g is known as the group velocity and is given from Eq. 2.29 by

$$v_g = \left. \frac{d\omega}{dk(\omega)} \right|_{\omega=\omega_0} \tag{2.30}$$

The group velocity represents the speed at which information is transferred from transmitter to receiver.

Recall that our analysis of wave propagation in a uniform unbounded medium showed that $k(\omega)$ was a linear function of frequency; that is

$$k(\omega) = \omega\sqrt{\mu\epsilon}$$

Thus for a plane wave propagating in such a material, phase and group velocities are equal and given by

$$v_p = v_g = 1/\sqrt{\mu\epsilon}$$

As will be shown in Chapter 4, when the transmission medium is a waveguide, $k(\omega)$ is no longer a linear function of frequency. In such instances it is useful to examine the so called ω-k diagram which graphs ω versus $k(\omega)$, shown schematically in Figure 2.12. As shown in the figure, the slope of a line drawn from the origin to the point on the curve located at carrier frequency, ω_0 gives the phase velocity. Similarity, the slope of a tangent to the curve at ω_0 yields the group velocity. It is important to note that as the carrier frequency is changed, both v_p and v_g will in general also change.

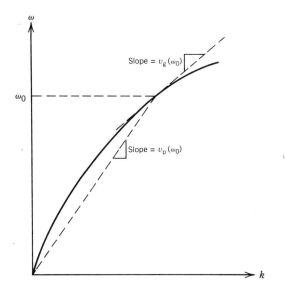

Figure 2.12 Graphical demonstration of the method for computing group velocity $v_g(\omega_0)$ and phase velocity $v_p(\omega_0)$.

Example 2.3

A plane wave propagating in a medium containing N ionized particles per unit volume of mass m and charge q can be shown to have a propagation constant given by

$$k(\omega) = \omega \sqrt{\mu_0 \epsilon(\omega)}$$

where

$$\epsilon(\omega) = \epsilon_0(1 - \omega_p^2/\omega^2)$$

and ω_p is known as the plasma frequency defined by

$$\omega_p = \sqrt{Nq^2/m\epsilon_0}$$

When the charged particles are electrons and the density N is in units of m^{-3}, then the plasma frequency is given by

$$\omega_p \cong 56.4\sqrt{N}$$

A plot of ω versus $k(\omega)$ is shown in Figure 2.13. Note from the graph that as ω approaches ω_p the phase velocity becomes infinite. Although this result may appear disturbing, recall that from a physical point of view, it is the group velocity or signal velocity which we must require to be less than the speed of light. We observe from the ω-k plot that as ω approaches ω_p, the group velocity in fact approaches zero.

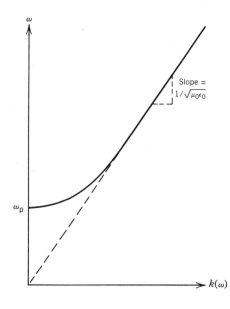

Figure 2.13 Dispersion diagram for an ionized plasma.

Additionally, it is observed that in the limit of very high frequency, phase and group velocities approach the constant value $1/\sqrt{\mu_0\epsilon_0}$.

2.7   POLARIZATION

In Section 2.4.4, it was shown that the electric field of a plane wave can have no component parallel to the direction of propagation in an isotropic media. Therefore, the most general form for $\underline{\mathbf{E}}$ is

$$\underline{\mathbf{E}}(z) = (\hat{x}\underline{E}_{x0} + \hat{y}\underline{E}_{y0})e^{-jk_z z}$$

Letting

$$\underline{E}_{x0} = E_{x0}e^{j\phi_x}$$

$$\underline{E}_{y0} = E_{y0}e^{j\phi_y}$$

then

$$\underline{\mathbf{E}}(z) = (\hat{x}E_{x0}e^{j\phi_x} + \hat{y}E_{y0}e^{j\phi_y})e^{-jk_z z}$$

$$= \underline{E}_{x0}(\hat{x} + \hat{y}Ae^{j\phi})e^{-jk_z z}$$

where

$$A \equiv E_{y0}/E_{x0}$$

$$\phi = \phi_y - \phi_x$$

Without loss of generality we set $\underline{E}_{x0} = 1$, so that an arbitrarily polarized wave is represented as

$$\underline{\mathbf{E}}(z) = (\hat{x} + \hat{y}Ae^{j\phi})e^{-jk_z z} \tag{2.31}$$

Let us examine some different types of polarization.

Case I. $A = 0$

$$\underline{\mathbf{E}}(z) = \hat{x}e^{-jk_z z}$$

and

$$\mathbf{E}(z, t) = \mathrm{Re}[\hat{x}e^{j(\omega t - k_z z)}]$$

$$= \hat{x}\cos(\omega t - k_z z)$$

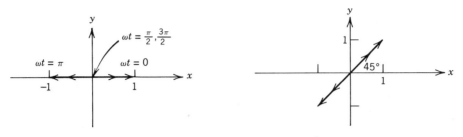

Figure 2.14 Linearly polarized wave.   **Figure 2.15** Linearly polarized wave at 45° angle.

The motion of the electric field vector at some fixed plane, say, $z = 0$, is shown in Figure 2.14. The electric field moves back and forth with time but always points along the $\hat{x}$ axis. This is known as a linearly polarized wave along $\hat{x}$.

Case II. $A = 1, \phi = 0$ (Figure 2.15)

$$\underline{\mathbf{E}}(z) = (\hat{x} + \hat{y})e^{-jk_z z}$$

$$\mathbf{E}(z, t) = (\hat{x} + \hat{y}) \cos(\omega t - k_z z)$$

Again we obtain a linearly polarized wave with the electric field vector pointing at 45° with respect to the x axis.

Case III. $A = 2, \phi = 0$ (Figure 2.16)

$$\mathbf{E}(z, t) = (\hat{x} + 2\hat{y}) \cos(\omega t - k_z z)$$

Case IV. $A = 1, \phi = \pi/2$

$$\underline{\mathbf{E}}(z) = (\hat{x} + j\hat{y})e^{-jk_z z}$$

$$\mathbf{E}(z, t) = \hat{x} \cos(\omega t - k_z z) - \hat{y} \sin(\omega t - k_z z)$$

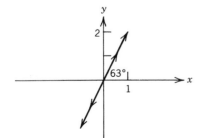

Figure 2.16 Still linearly polarized.

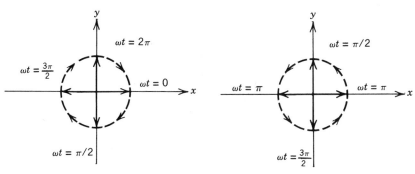

Figure 2.17 Left-hand circularly polarized.   **Figure 2.18** Right-hand circularly polarized.

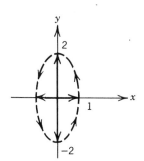

Figure 2.19 Right-hand elliptically polarized.

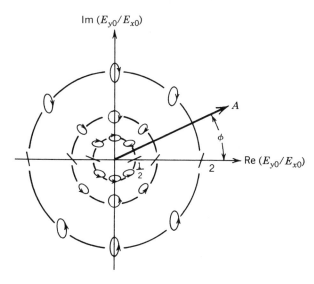

Figure 2.20 Polarization diagram. (After Kong (Ref. 4). Copyright © 1975, John Wiley & Sons. Reproduced by permission.)

The easiest way to determine **E** is to set $z = 0$ and plot a few points for different times (Figure 2.17).

Note that the tip of the electric field vector traces out a circle. The wave is defined to be left-hand circularly polarized. This convention is obtained by placing the fingers of the left hand along the perimeter of the circle in the direction of vector rotation and noting that the thumb then points in the direction of propagation ($+\hat{z}$).

Case V. $A = 1, \phi = -\pi/2$ (Figure 2.18)

This is similar to Case IV. We obtain

$$\mathbf{E}(z, t) = \hat{x} \cos(\omega t - k_z z) + \hat{y} \sin(\omega t - k_z z)$$

which is a right-hand circularly polarized wave.

Case VI. $A = 2, \phi = -\pi/2$

$$\mathbf{E}(z, t) = \hat{x} \cos(\omega t - k_z z) + \hat{y}2 \sin(\omega t - k_z z)$$

Note that $|E_x|^2 + \frac{1}{4} |E_y|^2 = 1$ which is the equation of an ellipse as shown in Figure 2.19. The wave is said to be right-hand elliptically polarized. For any general A and ϕ the type of polarization can be determined using the above techniques or by making use of the polarization chart shown in Figure 2.20. This chart indicates the nature of the polarization on the complex $\underline{E_{y0}/E_{x0}}$ plane. Note that on this plane $\underline{E_{y0}/E_{x0}}$ can be represented in polar coordinates by magnitude A and phase ϕ.

2.8   DUALITY

The concept of duality discussed in this section is particularly useful as a shortcut in obtaining solutions to a number of problems which are relevant to integrated optics. Let us consider Maxwell's equations in a source-free medium characterized by μ and ϵ:

$$\nabla \times \underline{\mathbf{E}} = -j\omega\mu\underline{\mathbf{H}}$$

$$\nabla \times \underline{\mathbf{H}} = j\omega\epsilon\underline{\mathbf{E}}$$

$$\nabla \cdot \epsilon\underline{\mathbf{E}} = 0$$

$$\nabla \cdot \mu\underline{\mathbf{H}} = 0$$

The high degree of symmetry between **E** and **H** is clearly visible by inspection of these four equations. In fact, note that if the following interchanges are made

$$\underline{E} \longrightarrow -\underline{H}$$

$$\underline{H} \longrightarrow \underline{E}$$

$$\mu \longrightarrow \epsilon$$

$$\epsilon \longrightarrow \mu$$

we obtain the identical equations. This implies that if one solution to Maxwell's equations has been obtained for some particular problem, a second solution is automatically obtained by the interchange indicated above since it satisfies the identical set of equations.

Example 2.4

Consider an $\hat{x}$-polarized plane wave propagating in the $\hat{z}$ direction as shown in Figure 2.21. Obtain the dual solution to this problem.

Solution

The wave takes the form

$$\left.\begin{array}{l} E = \hat{x} \\ H = \hat{y}\, \dfrac{1}{\eta} \end{array}\right\} e^{-jk_z z}$$

where

$$k_z = \omega\sqrt{\mu\epsilon}$$

$$\eta = \sqrt{\mu/\epsilon}$$

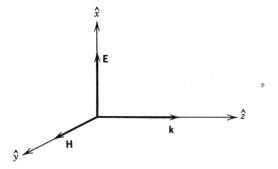

Figure 2.21 An $\hat{x}$-polarized plane wave propagating along z.

Duality requires that a second solution is given by

$$\left.\begin{array}{l} \mathbf{H} = -\hat{x} \\ \mathbf{E} = \hat{y}\eta \end{array}\right\} e^{-jk_z z}$$

which is simply a plane wave polarized orthogonally with respect to the original solution.

2.9 PLANE-WAVE SPECTRUM OF FINITE-WIDTH BEAMS

While the properties of the unbounded plane wave are particularly easy to analyze, in practice all true electromagnetic radiation is generated from a source of finite dimensions and must, therefore, be of finite spatial extent. True electromagnetic radiation can, therefore, only approximate the ideal plane wave. For example, a light beam from a He–Ne laser, or for that matter from a flashlight, does not constitute a plane wave; yet on intuitive grounds we would expect both to obey closely the laws of transmission and reflection for true plane waves. Clearly, as the beam width is increased such a wave should approximate more closely the ideal plane wave. Because optical beams of finite width will prove useful in much of the interpretation of guided-wave and coupling theory to be derived, the relationship between the infinite plane wave and finite-width beam is analyzed in this section.

Let us consider for simplicity the two-dimensional problem of an optical beam propagating along $\hat{z}$ that is uniform in the $\hat{y}$ direction and has beam width W in the $\hat{x}$ direction at the plane $z = 0$, as shown in Figure 2.22. The electric field associated with the beam is assumed to be $\hat{y}$ polarized and of unity amplitude.

We note that the beam must be a solution to Maxwell's equations and, therefore, must also satisfy the wave equation. It has already been shown that the solutions to this equation, in absence of any y variation and assuming $\hat{y}$ polarization, are of the form

$$\mathbf{E}(x, y) = \hat{y}E e^{-jk_x x} e^{-jk_z z}$$

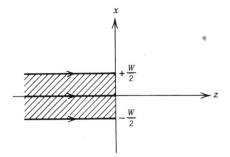

Figure 2.22 Two-dimensional optical beam of width W at $z = 0$.

with

$$k_x^2 + k_z^2 = \omega^2 \mu_0 \epsilon_0 = k_0^2$$

Here the wave amplitude E, and propagation direction are arbitrary. Note that the propagation direction can be specified by giving a value to the wavenumber k_x since k_z is then determined from the above plane-wave dispersion relation

$$k_z = \sqrt{k_0^2 - k_x^2} \tag{2.32}$$

The amplitude and direction of any arbitrary propagating plane-wave solution can, therefore, be specified by a single variable, $E(k_x)$. Therefore, the most general solution to the wave equation is a superposition of plane waves propagating in all directions with the amplitude of each component given by $E(k_x)$. That is

$$\mathbf{E}(x, z) = \hat{y} \int_{-\infty}^{\infty} dk_x \, E(k_x) e^{-jk_x x} e^{-jk_z z}$$

Note that in integrating over k_x, the values of k_z corresponding to $k_x > k_0$ will be imaginary, representing plane-wave components which are exponentially decaying in the direction of propagation.

If it is possible for such a superposition of waves to represent the optical beam then it is necessary for any x and z that

$$E_y(x, z) = \int_{-\infty}^{\infty} dk_x \, E(k_x) e^{-jk_x x} e^{-jk_z z}$$

In particular, at $z = 0$

$$E_y(x, z = 0) = \int_{-\infty}^{\infty} dk_x \, E(k_x) e^{-jk_x x}$$

Thus, *it is observed that the plane-wave weighting coefficients $E(k_x)$ are given simply by the Fourier transform of the beam transverse distribution at $z = 0$.* For the case shown in Figure 2.22

$$E_y(x, 0) = 1 \qquad |x| < W/2$$

$$= 0 \qquad \text{otherwise}$$

so that

$$E(k_x) = \frac{1}{2\pi} \int_{-\infty}^{\infty} E_y(x, 0) \, e^{jk_x x}$$

$$= \frac{1}{\pi} \frac{W}{2} \frac{\sin k_x W/2}{k_x W/2}$$

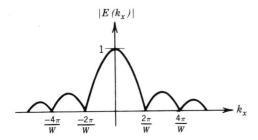

Figure 2.23 Plane-wave spectrum for beam of width W.

which is plotted in Figure 2.23. Note that the "bandwidth" of $E(k_x)$ is limited by

$$|(k_x)_{max}| \cong 2\pi/W$$

If the magnitude of $(k_x)_{max}$ is much smaller than k_0 then from Eq. 2.32 $k_z \cong k_0$ and the plane waves that make up the beam have significant amplitude only over a maximum angular spread θ_{max} given by

$$\tan(\theta_{max}) = \frac{(k_x)_{max}}{k_z} \cong \frac{2\pi/W}{k_0}$$

or

$$\theta_{max} \cong \frac{2\pi/W}{2\pi/\lambda_0} = \frac{\lambda_0}{W}$$

Thus, the beam angular spread is $\Delta\theta \cong 2\theta_{max}$, which is given by

$$\Delta\theta \cong \frac{2\lambda_0}{W} \tag{2.33}$$

as shown in Figure 2.24. Provided that the width of the beam is many wavelengths then its angular spread is small and it will remain well collimated over large distances. The distance z_c over which the beam remains well collimated is seen from the geometry of Figure 2.24 to be given by

$$(W/2)/z_c \cong \Delta\theta/2$$

Substituting for $\Delta\theta$ from Eq. 2.33 then yields

$$z_c \cong \frac{W^2}{2\lambda_0} \tag{2.34}$$

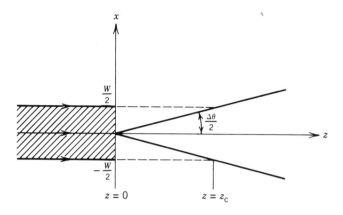

Figure 2.24 Diffraction by a beam of width W. Beam remains collimated over the distance z_c.

2.10 POWER FLOW

2.10.1 Poynting's Theorem

In this section we analyze how the various electromagnetic field terms are interrelated in terms of power flow. These variables are connected via Maxwell's curl equations, which for an isotropic medium are given by

$$\nabla \times \mathbf{E} = -\mu \frac{\partial \mathbf{H}}{\partial t} \tag{2.35}$$

$$\nabla \times \mathbf{H} = \epsilon \frac{\partial \mathbf{E}}{\partial t} + \mathbf{J} \tag{2.36}$$

In general, the current density $\mathbf{J}$ can be separated into two components:

$$\mathbf{J} = \mathbf{J}_s + \mathbf{J}_c$$

The term $\mathbf{J}_s$ represents any current sources that give rise to electromagnetic fields, whereas $\mathbf{J}_c = \sigma \mathbf{E}$ represents any conduction currents induced by the presence of electric fields. As an example, an antenna shown in Figure 2.25 is excited with an alternating current source having current density $\mathbf{J}_s$ and gives rise to electromagnetic fields that radiate away. If a region of finite conductivity exists, then the presence of the electric field in this region will induce conduction currents $\mathbf{J}_c$ to flow.

Notice that since $\mathbf{E}$ has units of volts per meter and $\mathbf{J}$ has units of amperes per meter squared, then the quantity $\mathbf{E} \cdot \mathbf{J}$ has the dimension of power per unit volume (watts per cubic meter). From Eq. 2.36 above we have, therefore,

$$\mathbf{E} \cdot \mathbf{J} = \mathbf{E} \cdot \nabla \times \mathbf{H} - \epsilon \mathbf{E} \cdot \frac{\partial \mathbf{E}}{\partial t}$$

Using the vector identity

$$\nabla \cdot (\mathbf{A} \times \mathbf{B}) = \mathbf{B} \cdot \nabla \times \mathbf{A} - \mathbf{A} \cdot \nabla \times \mathbf{B}$$

then

$$\mathbf{E} \cdot \mathbf{J} = \mathbf{H} \cdot \nabla \times \mathbf{E} - \nabla \cdot (\mathbf{E} \times \mathbf{H}) - \epsilon \mathbf{E} \cdot \frac{\partial \mathbf{E}}{\partial t}$$

Substitution of Eq. 2.35 for the curl of $\mathbf{E}$ into the above expression yields

$$\mathbf{E} \cdot \mathbf{J} = - \mu \mathbf{H} \cdot \frac{\partial \mathbf{H}}{\partial t} - \nabla \cdot (\mathbf{E} \times \mathbf{H}) - \epsilon \mathbf{E} \cdot \frac{\partial \mathbf{E}}{\partial t}$$

Noting that

$$\frac{1}{2} \frac{\partial}{\partial t} |\mathbf{A}|^2 = \frac{1}{2} \frac{\partial}{\partial t} \mathbf{A} \cdot \mathbf{A} = \frac{1}{2} \left(\mathbf{A} \cdot \frac{\partial \mathbf{A}}{\partial t} + \frac{\partial \mathbf{A}}{\partial t} \cdot \mathbf{A} \right) = \mathbf{A} \cdot \frac{\partial \mathbf{A}}{\partial t}$$

then

$$\mathbf{E} \cdot \mathbf{J} = - \frac{\mu}{2} \frac{\partial |\mathbf{H}|^2}{\partial t} - \frac{\epsilon}{2} \frac{\partial |\mathbf{E}|^2}{\partial t} - \nabla \cdot (\mathbf{E} \times \mathbf{H})$$

Because $\mathbf{E} \cdot \mathbf{J}$ has units of power per unit volume rather than power we integrate all quantities over an arbitrary closed volume V as shown in Figure 2.25. Let the surface bounding V be S and an outward unit normal to the surface be $\hat{n}$ as shown in the figure. Then we obtain

$$- \iiint_V dv \, \mathbf{E} \cdot \mathbf{J} = \frac{\partial}{\partial t} \iiint_V dv \, (\tfrac{1}{2} \mu |\mathbf{H}|^2 + \tfrac{1}{2} \epsilon |\mathbf{E}|^2) + \oiint_S ds \, \hat{n} \cdot (\mathbf{E} \times \mathbf{H}) \quad (2.37)$$

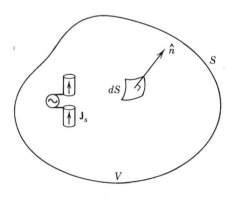

Figure 2.25 Volume V over which power balance is to be made.

where we have made use of the identity

$$\iiint_V dv\, \nabla \cdot \mathbf{A} = \oiint_S ds\, \hat{n} \cdot \mathbf{A}$$

Equation 2.37 is a statement of Poynting's theorem and its various terms are identified using energy-conservation arguments for electromagnetic fields. The terms $\frac{1}{2}\mu|\mathbf{H}|^2$ and $\frac{1}{2}\epsilon|\mathbf{E}|^2$ have dimensions of energy per unit volume and are interpreted as in electrostatics to be the instantaneous energy densities stored in magnetic and electric fields, respectively. Thus, the term

$$\frac{\partial}{\partial t}\iiint_V dv\, (\tfrac{1}{2}\mu|\mathbf{H}|^2 + \tfrac{1}{2}\epsilon|\mathbf{E}|^2)$$

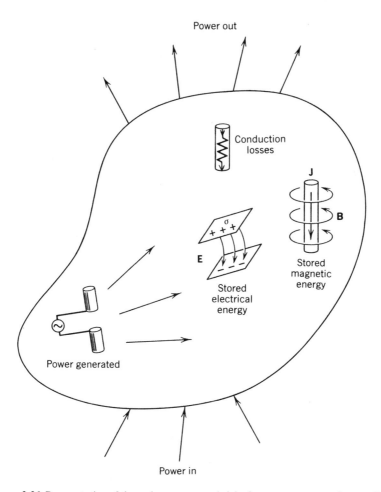

Figure 2.26 Demonstration of the various sources and sinks for energy storage and power flow.

represents the rate of increase of stored energy in volume V. The term

$$
-\int\!\!\int\!\!\int_V dv\ \mathbf{E} \cdot \mathbf{J} = \overbrace{-\int\!\!\int\!\!\int_V dv\ \mathbf{E} \cdot \mathbf{J}_s}^{\text{Power generated}} - \overbrace{\int\!\!\int\!\!\int_V dv\ \sigma|\mathbf{E}|^2}^{\text{Power dissipated}}
$$

is interpreted as the net power generated in the volume V due to source $\mathbf{J}_s$ minus the power dissipated due to conduction losses resulting from finite σ. Finally the term,

$$
\oiint_S ds\ \hat{n} \cdot (\mathbf{E} \times \mathbf{H})
$$

is interpreted as the power flowing out of the volume V. The quantity

$$
\mathbf{S} \equiv \mathbf{E} \times \mathbf{H}
$$

is called the Poynting vector and represents the magnitude and direction of the power flux at any point. It has units of power per unit area. Poynting's theorem is, therefore, interpreted as follows: The power generated within an arbitrary volume V minus conduction losses is equal to the time rate of increase of stored energy within V plus the net power flowing out of V through the surface S. This relation is shown schematically in Figure 2.26.

Example 2.5

A plane wave propagating in the $\hat{z}$ direction has the following fields:

$$
\mathbf{E}(\mathbf{r}, t) = \hat{x} E_{y0} \cos(\omega t - kz)
$$

$$
\mathbf{H}(\mathbf{r}, t) = \hat{y}\frac{E_{y0}}{\eta} \cos(\omega t - kz) \qquad \eta = \sqrt{\mu/\epsilon}
$$

(a) Compute the Poynting vector $\mathbf{S}$ and stored electric and magnetic energy densities W_e and W_m.
(b) Assuming a rectangular volume V of dimensions $x = a$, $y = b$, and $z = c$, verify that Poynting's theorem is satisfied.

Solution

(a)
$$
\mathbf{S} = \mathbf{E} \times \mathbf{H}
$$

$$
= (\hat{x} \times \hat{y})\frac{E_{y0}^2}{\eta} \cos^2(\omega t - kz)
$$

$$
= \hat{z}\frac{E_{y0}^2}{\eta} \cos^2(\omega t - kz)
$$

$$W_e = \tfrac{1}{2}\epsilon|\mathbf{E}|^2 = \frac{\epsilon}{2} E_{y0}^2 \cos^2(\omega t - kz)$$

$$W_m = \tfrac{1}{2}\mu|\mathbf{H}|^2 = \frac{\mu}{2}\left(\frac{E_{y0}}{\eta}\right)^2 \cos^2(\omega t - kz)$$

(b) The volume V is shown in Figure 2.27. Since there are no sources or losses, the left side of Eq. 2.37 is zero. Let us first compute the total Poynting power flowing through the surface S of the volume V. Since $\mathbf{S}$ is $\hat{z}$ directed, the only contributions will be to the surfaces defined by $z = 0$ and $z = c$. Thus,

$$\oint_S ds \, \hat{n} \cdot \mathbf{S} = \left[-\int_0^a dx \int_0^b dy \, S_z(z = 0) + \int_0^a dx \int_0^b dy \, S_z (z = c) \right]$$

$$= -\frac{E_{y0}^2}{\eta} ab[\cos^2 \omega t - \cos^2 (\omega t - kc)]$$

$$= -\frac{E_{y0}^2}{2\eta} ab[\cos 2\omega t - \cos 2(\omega t - kc)]$$

The total stored energy within V is

$$\iiint_V dv(W_e + W_m)$$

$$= \frac{E_{y0}^2}{2} \int_0^a dx \int_0^b dy \int_0^c dz\left(\epsilon + \frac{\mu}{\eta^2}\right) \cos^2(\omega t - kz)$$

$$= \frac{\epsilon E_{y0}^2}{2} ab\int_0^c dz[1 + \cos 2(\omega t - kz)]$$

Taking the time derivative to obtain the rate of increase of stored energy eliminates the first term within the brackets and yields for the second term

$$\frac{\partial}{\partial t} \iiint_V dv(W_e + W_m)$$

$$= -\epsilon E_{y0}^2 \frac{ab\omega}{2k}[\cos 2(\omega t - kc) - \cos 2\omega t]$$

$$= \frac{E_{y0}^2}{2\eta} ab[\cos 2\omega t - \cos 2(\omega t - kc)]$$

where we have used the fact that $k = \omega\sqrt{\mu\epsilon}$. Comparing the above result with the total outward flow of Poynting power shows that Poynting's theorem is indeed satisfied.

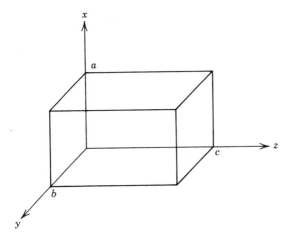

Figure 2.27 Volume used for Example 2.5.

2.10.2 Complex Poynting's Theorem

It has been demonstrated that representing time-harmonic fields in terms of complex phasor amplitudes is often convenient because of the elimination of time as a variable. Care must be taken, however, in the representation of power quantities in terms of these phasors because they contain products of two complex quantities. In particular, we note that if $A(t)$ and $B(t)$ are two time-harmonic quantities

$$A(t) = \text{Re}(\underline{A}e^{j\omega t}) = \text{Re}(Ae^{j\phi_A}e^{j\omega t})$$

$$B(t) = \text{Re}(\underline{B}e^{j\omega t}) = \text{Re}(Be^{j\phi_B}e^{j\omega t})$$

then

$$A(t)B(t) = \text{Re}(\underline{A}e^{j\omega t})\, \text{Re}(\underline{B}e^{j\omega t}) \neq \text{Re}(\underline{AB}e^{j\omega t})$$

Thus, we cannot represent the product of two time-harmonic signals as the product of their phasors.

However, a simple relationship may be demonstrated between the time average of the product $A(t) B(t)$ and the phasors $\underline{A}$ and $\underline{B}$. Note that

$$A(t)B(t) = AB\cos(\omega t + \phi_A)\cos(\omega t + \phi_B)$$

$$= \frac{AB}{2}[\cos(\phi_A - \phi_B) + \cos(2\omega t + \phi_A + \phi_B)]$$

Taking the time average over one period yields

$$\langle A(t)B(t)\rangle = \frac{AB}{2}\cos(\phi_A - \phi_B)$$

where the brackets indicate the time average. But the right side of the above equality is given simply by

$$\tfrac{1}{2}\,\mathrm{Re}(A\underline{B}^*)$$

where B^* is the complex conjugate of B. We conclude that in general the time average of the product of two time-harmonic quantities is given by

$$\langle A(t)B(t)\rangle = \tfrac{1}{2}\,\mathrm{Re}(A\underline{B}^*) \tag{2.38}$$

The above result can be extended to vector products since these may be represented as the sum of scalar products. Therefore, it follows, for example, that the time-average Poynting power density $\langle S \rangle$ is given by

$$\langle S \rangle = \tfrac{1}{2}\,\mathrm{Re}(\underline{\mathbf{E}} \times \underline{\mathbf{H}}^*) \tag{2.39}$$

where the quantity

$$\underline{\mathbf{S}} = \underline{\mathbf{E}} \times \underline{\mathbf{H}}^* \tag{2.40}$$

is defined as the complex Poynting vector. To interpret the meaning of the time-average Poynting power we reformulate Poynting's theorem in terms of complex field amplitudes. From Eqs. 2.13 and 2.14

$$\nabla \times \underline{\mathbf{E}} = -\,j\omega\mu\underline{\mathbf{H}} \tag{2.41}$$

$$\nabla \times \underline{\mathbf{H}} = j\omega\epsilon\underline{\mathbf{E}} + \underline{\mathbf{J}} \tag{2.42}$$

Taking $\underline{\mathbf{H}}^* \cdot$ (Eq. 2.41) $-\ \underline{\mathbf{E}} \cdot$ (Eq. 2.42)* yields

$$\underline{\mathbf{H}}^* \cdot \nabla \times \underline{\mathbf{E}} - \underline{\mathbf{E}} \cdot \nabla \times \underline{\mathbf{H}}^* = -\,j\omega\mu|\underline{\mathbf{H}}|^2 + j\omega\epsilon|\underline{\mathbf{E}}|^2 - \underline{\mathbf{E}} \cdot \underline{\mathbf{J}}^*$$

or

$$\nabla \cdot (\underline{\mathbf{E}} \times \underline{\mathbf{H}}^*) = -\,j\omega(\mu|\underline{\mathbf{H}}|^2 - \epsilon|\underline{\mathbf{E}}|^2) - \underline{\mathbf{E}} \cdot \underline{\mathbf{J}}^*$$

In integral form this becomes

$$\frac{1}{2}\oiint_S ds\,\hat{n} \cdot \underline{\mathbf{S}} = -\,2j\omega\iiint_V dv(\tfrac{1}{4}\mu|\underline{\mathbf{H}}|^2 - \tfrac{1}{4}\epsilon|\underline{\mathbf{E}}|^2) - \frac{1}{2}\iiint_V dv\,\underline{\mathbf{E}} \cdot \underline{\mathbf{J}}^* \tag{2.43}$$

which is a statement of conservation of complex power. The left side of Eq. 2.43 represents the total complex power flowing out of the closed surface S. We examine the significance of the various terms in the above expression by equating real and imaginary parts. The real part of Eq. 2.43 yields

$$\frac{1}{2} \oiint_S ds\, \hat{n} \cdot \text{Re}(\underline{\mathbf{S}}) = \oiint_S ds\, \hat{n} \cdot \langle \mathbf{S} \rangle = P_s - P_c \tag{2.44}$$

where

$$P_s \equiv -\frac{1}{2} \iiint_V dv\, \text{Re}(\underline{\mathbf{E}} \cdot \underline{\mathbf{J}}_s^*)$$

and

$$P_c \equiv \frac{1}{2} \iiint_V dv\, \sigma|\underline{\mathbf{E}}|^2$$

P_s and P_c are interpreted, respectively, as the total real or time-average power generated by source $\mathbf{J}_s$ in volume V and the total power dissipated in V due to conduction losses. Thus, the total flux of time-average Poynting power out of volume V is seen equal to the difference between time-average power generated and power dissipated.

Similarly, equating imaginary parts yields

$$\frac{1}{2} \oiint_S ds\, \hat{n} \cdot \text{Im}(\underline{\mathbf{S}}) = -2\omega(W_m - W_e) + P_r \tag{2.45}$$

where

$$W_m \equiv \iiint_V dv(\tfrac{1}{4}\mu|\underline{\mathbf{H}}|^2)$$

$$W_e \equiv \iiint_V dv(\tfrac{1}{4}\epsilon|\underline{\mathbf{E}}|^2)$$

and

$$P_r \equiv -\frac{1}{2} \iiint_V dv\, \text{Im}\,(\underline{\mathbf{E}} \cdot \underline{\mathbf{J}}_s^*)$$

The terms W_m and W_e are interpreted as total magnetic and electric energy stored in the volume V in analogy with their electrostatic counterparts. P_r is interpreted in analogy to circuit theory as the reactive power delivered to the volume V by source $\mathbf{J}_s$. Thus, Eq. 2.45 states that the total outward flux of the imaginary part of the Poynting vector through a closed surface S is equal to the reactive power supplied minus 2ω times the difference between magnetic and electric stored energy.

Example 2.6

Show that the complex form of Poynting's theorem is satisfied for the plane wave and volume V described in Example 2.5.

Solution

The complex electric and magnetic fields are

$$\underline{\mathbf{E}}(\mathbf{r}) = \hat{x} E_{y0} \, e^{-jkz}$$

$$\underline{\mathbf{H}}(\mathbf{r}) = \hat{y} \frac{E_{y0}}{\eta} \, e^{-jkz}$$

The complex Poynting vector is

$$\underline{\mathbf{S}} = \underline{\mathbf{E}} \times \underline{\mathbf{H}}^* = \hat{z} \frac{E_{y0}^2}{\eta}$$

Note that for a plane wave $\underline{\mathbf{E}}$ and $\underline{\mathbf{H}}$ are in phase and thus $\underline{\mathbf{S}}$ is purely real.

Since there are no sources or loss terms, the real part of Poynting's theorem requires that

$$\tfrac{1}{2} \operatorname{Re} \oiint_S ds \, \hat{n} \cdot \underline{\mathbf{S}} = 0$$

Integrating over the surface of V yields

$$\tfrac{1}{2} \operatorname{Re} \left[\int_0^a dx \int_0^b dy \left(-\frac{E_{y0}^2}{\eta} \right) + \int_0^a dx \int_0^b dy \left(\frac{E_{y0}^2}{\eta} \right) \right] = 0$$

so that the real part of Poynting's theorem is satisfied.

From Eq. 2.45, since $\operatorname{Im}(\underline{\mathbf{S}}) = 0$ we must show that $W_m = W_e$. Now

$$W_e = \frac{1}{4} \int_0^a dx \int_0^b dy \int_0^c dz \, \epsilon E_{y0}^2 = abc \, \epsilon E_{y0}^2 / 4$$

and

$$W_m = \frac{1}{4} \int_0^a dx \int_0^b dy \int_0^c dz \, \mu \left(\frac{E_{y0}}{\eta} \right)^2 = \frac{abc}{4} \mu \frac{E_{y0}^2}{\mu/\epsilon} = W_e$$

Thus the imaginary part of Poynting's theorem is also satisfied.

PROBLEMS

2.1 Determine if the set of fields listed below satisfy Maxwell's equations 2.1–2.4 in a source-free region of free space.

$$E(\mathbf{r}, t) = \hat{x} \cos(\omega t - k_0 z) + 2\hat{y} \sin(\omega t - k_0 z)$$

$$H(\mathbf{r}, t) = \frac{k_0}{\omega \mu_0} [2\hat{x} \sin(\omega t - k_0 z) - \hat{y} \cos(\omega t - k_0 z)]$$

where

$$k_0^2 = \omega^2 \mu_0 \epsilon_0$$

2.2 Repeat problem 2.1 for the following fields:

$$E(\mathbf{r}, t) = \hat{y} \cos(\omega t - k_x x - k_z z)$$

$$H(\mathbf{r}, t) = \frac{1}{\omega \mu_0} (-\hat{x} k_z + \hat{z} k_x) \cos(\omega t - k_x x - k_z z)$$

where

$$k_x^2 + k_z^2 = k_0^2$$

2.3 Represent the following fields in phasor form:
(a) $E(\mathbf{r}, t) = 2\hat{y} \sin(\omega t - kz)$
(b) $E(\mathbf{r}, t) = 2\hat{x} \sin(\omega t - kz) - \hat{y} \cos(\omega t - kz)$
(c) $E(\mathbf{r}, t) = (-\hat{x} k_z + \hat{z} k_x) \cos(\omega t - k_x x - k_z z)$
(d) $E(\mathbf{r}, t) = \hat{y} \cos(\omega t - k_z z) e^{-\alpha_r x}$

2.4 (a) Give the true time–space representations of the following complex electric fields:
   (1) $E(\mathbf{r}) = 3\hat{x} e^{-jkz}$
   (2) $\overline{E}(\mathbf{r}) = (\hat{x} + 2j\hat{y}) e^{-jkz}$
   (3) $\overline{E}(\mathbf{r}) = 2(\hat{x} - j\hat{y}) e^{-jkz}$
   (b) What type of polarization is associated with each of the above fields?

2.5 A plane wave propagating along the z direction has an electric field that in complex form is given by

$$E(\mathbf{r}) = 2\hat{y} e^{-jk_z z}$$

(a) Sketch at $t = 0$ the amplitude of the electric and magnetic fields $E(\mathbf{r}, t = 0)$, $H(\mathbf{r}, t = 0)$ as a function of z.
(b) Repeat for $\omega t = \pi/2$.

2.6 Using Eq. 2.21, compute the complex magnetic fields $\mathbf{H(r)}$ for plane waves propagating in free space and having the following electric fields:

(a) $\mathbf{E(r)} = \hat{y}e^{-jkz}$

(b) $\overline{\mathbf{E}}(\mathbf{r}) = \hat{y}e^{-j(k_x x + k_z z)}$

(c) $\overline{\mathbf{E}}(\mathbf{r}) = (\hat{x}k_z - \hat{z}k_x)e^{-j(k_x x + k_z z)}$

(d) $\overline{\mathbf{E}}(\mathbf{r}) = 2\hat{y}\cos(k_x x)e^{-jk_z z}$

2.7 (a) Show that the following electric field can be represented as the superposition of four propagating plane waves:

$$\underline{\mathbf{E}}(\mathbf{r}) = \hat{y}\cos(k_x x)\sin(k_z z)$$

where $k_x^2 + k_z^2 = \omega^2 \mu_0 \epsilon_0$.

(b) Compute the corresponding magnetic field $\underline{\mathbf{H}}(\mathbf{r})$.

2.8 A signal is to be sent through the atmosphere that contains $N = 1 \times 10^{17}$ ionized electrons per cubic meter.

(a) Give a general expression for the phase and group velocities.

(b) Suppose the signal is centered about a carrier frequency $f_0 = 12$ GHz and has a bandwidth $\Delta f = 2$ MHz. What is the group and phase velocity at the center frequency f_0? If the receiver is 10 km away, what is the difference in arrival time for the frequency components of the signal at $f = f_0 - \Delta f$ and $f = f_0 + \Delta f$?

2.9 What is the minimum slit size W through which a He–Ne laser beam ($\lambda_0 = 0.6328$ μm) can be passed so that the beam is well collimated on a screen located 10 cm away?

2.10 A time-varying current source $i_s(t)$ supplies power to the RLC circuit shown in Figure 2.28:

Derive Poynting's theorem for the circuit above in terms of R, L, C, $i_s(t)$, $v_s(t)$, and $v_c(t)$. Identify the source, dissipation and energy storage terms.

2.11 Give the complex form of Poynting's theorem for the circuit in problem 2.10 when $i_s(t)$ is sinusoidal. Identify all source, dissipation, and storage terms in your expression.

2.12 Prove that for time-harmonic fields, Maxwell's equation $\nabla \cdot \mathbf{D}(\mathbf{r}, t) = \rho(\mathbf{r}, t)$ can be written as $\nabla \cdot \underline{\mathbf{D}}(\mathbf{r}) = \underline{\rho}(\mathbf{r})$.

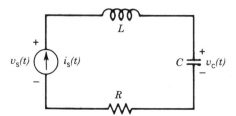

Figure 2.28

2.13   A time-harmonic electromagnetic wave propagating in free space has an electric field with the following complex amplitude:

$$\underline{\mathbf{E}}(\mathbf{r}) = \hat{y}\,\sin(k_x x)e^{-jk_z z}$$

where k_x and k_z are both real.
(a) Find the complex representation for the magnetic field $\underline{\mathbf{H}}(\mathbf{r})$.
(b) Find the complex Poynting vector $\underline{\mathbf{S}}(\mathbf{r})$.
(c) What is the direction and magnitude of the time-average Poynting power?

2.14   (a) Show that the linearly polarized wave,

$$\underline{\mathbf{E}} = \hat{x}e^{-jkz}$$

can be represented as the superposition of two oppositely rotating circularly polarized waves.
(b) Show that the circularly polarized wave,

$$\underline{\mathbf{E}} = (\hat{x} + j\hat{y})e^{-jkz}$$

can be represented as the sum of two linearly polarized waves.
(c) Show that the arbitrarily polarized wave,

$$\underline{\mathbf{E}} = (a\hat{x}e^{j\phi_x} + b\hat{y}e^{j\phi_y})e^{-jkz}$$

can be represented as the sum of two oppositely rotating circularly polarized waves.

2.15   Consider a $\hat{z}$-propagating wave of the form

$$\underline{\mathbf{E}}(z) = (\hat{x}E_x e^{j\phi_x} + \hat{y}E_y e^{j\phi_y})e^{-jkz}$$

where $E_x = 1$, $E_y = 2$, $\phi_x = \pi/4$, $\phi_y = \pi/2$. Plot the path of the tip of the $\mathbf{E}$ vector for $z = 0$ as a function of time.

REFERENCES

1. Goodman, J. W. *Introduction to Fourier Optics.* New York: McGraw-Hill, 1968.

2. Harrington, R. F. *Time-Harmonic Electromagnetic Fields.* New York: McGraw-Hill, 1961.

3. Jordan, E. C., and Balmain, K. G. *Electromagnetic Waves and Radiating Systems.* 2nd ed. Englewood Cliffs, N.J.: Prentice-Hall, 1968.

4. Kong, J. A. *Theory of Electromagnetic Waves.* New York: Wiley, 1975.

5. Paul, C. R., and Nasar, S. A. *Introduction to Electromagnetic Fields.* New York: McGraw-Hill, 1982.

CHAPTER 3

Reflection and Transmission at a Dielectric Interface

3.1 INTRODUCTION

Any realizable integrated-optics component, whether it be a modulator, directional coupler, waveguide, etc., must have physical dimensions of finite extent. In terms of the component's optical properties, this same geometry is described through the specification of the variation in dielectric constant or refractive index as a function of spatial coordinates. To understand how a component operates, we must, therefore, know how spatial variation in dielectric constant modifies the properties of the optical radiation within the device.

Perhaps the simplest variation in permittivity that can be envisioned is the stepwise discontinuous jump exhibited between two different dielectric half spaces. Analysis of the effect of such an interface on an incident plane wave will provide significant insight into the behavior of more complex dielectric geometries and will act as an optical building block to facilitate their analysis.

3.2 BOUNDARY CONDITIONS BETWEEN TWO DIELECTRIC INTERFACES

To obtain a description of the fields in each dielectric half space shown in Figure 3.1, it is necessary to obtain the boundary conditions that relate these fields across the interface. These relations are obtained directly from the integral form of Maxwell's curl equations for a source-free region:

$$\oint_C \mathbf{E} \cdot \hat{l} \, dl = -\partial/\partial t \oiint_S \mathbf{B} \cdot \hat{n} \, dS \tag{3.1}$$

$$\oint_C \mathbf{H} \cdot \hat{l} \, dl = \partial/\partial t \oiint_S \mathbf{D} \cdot \hat{n} \, dS \tag{3.2}$$

Here the closed line integral along C bounds the surface S and $\hat{n}$ is the normal to the differential surface element dS as shown in Figure 3.1.

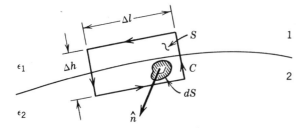

Figure 3.1 Geometry for computation of boundary conditions between regions 1 and 2.

If the limit is taken as Δh approaches zero while Δl is kept small but finite then, provided the fields **B** and **D** are finite, the right sides of Eqs. 3.1 and 3.2 approach zero while $\mathbf{E} \cdot \hat{l}$ and $\mathbf{H} \cdot \hat{l}$ approach their tangential values $\mathbf{E}_{\text{tan}}$ and $\mathbf{H}_{\text{tan}}$ at the interface. Since Δl is arbitrarily small, $\mathbf{E}_{\text{tan}}$ and $\mathbf{H}_{\text{tan}}$ can be approximated as constant over the range of integration yielding,

$$(\mathbf{E}_{\text{tan}})_1 \, \Delta l - (\mathbf{E}_{\text{tan}})_2 \, \Delta l = 0$$

$$(\mathbf{H}_{\text{tan}})_1 \, \Delta l - (\mathbf{H}_{\text{tan}})_2 \, \Delta l = 0$$

Since the location of our small surface S along the interface was arbitrary, these relations must hold at all points along the boundary. Thus, *we conclude that the tangential components of E and H must be continuous across the dielectric interface at all points along the boundary.*

3.3 PHASE MATCHING AND SNELL'S LAW

Let us consider a plane wave propagating through a half space characterized by ϵ_1 and μ_1, region 1, and incident upon an interface that separates region 1 from half-space region 2, characterized by ϵ_2 and μ_2, as shown in Figure 3.2. Let the amplitude of incident, reflected, and transmitted-wave electric fields be given by $\mathbf{A}_i$, $\mathbf{A}_r$, and $\mathbf{A}_t$ and

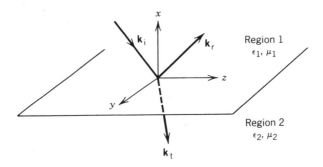

Figure 3.2 Plane wave incident from region 1 upon region 2.

the corresponding wave vectors by $\mathbf{k}_i$, $\mathbf{k}_r$, and $\mathbf{k}_t$. Thus,

$$\mathbf{E}_i(\mathbf{r}) = \mathbf{A}_i \, e^{-j(\mathbf{k}_i \cdot \mathbf{r})}$$

$$\mathbf{E}_r(\mathbf{r}) = \mathbf{A}_r \, e^{-j(\mathbf{k}_r \cdot \mathbf{r})}$$

$$\mathbf{E}_t(\mathbf{r}) = \mathbf{A}_t \, e^{-j(\mathbf{k}_t \cdot \mathbf{r})}$$

We have omitted the underbar below complex quantities for convenience and the complex nature of these quantities will be implicit throughout the remainder of the text.

Boundary conditions require that the tangential electric field is continuous across the boundary at $x = 0$. Thus,

$$[\mathbf{E}_i(0^+, y, z) + \mathbf{E}_r(0^+, y, z)]_{\text{tan}} = [\mathbf{E}_t(0^-, y, z)]_{\text{tan}}$$

which implies that

$$[\mathbf{A}_i e^{-jk_{iy}y} e^{-jk_{iz}z} + \mathbf{A}_r e^{-jk_{ry}y} e^{-jk_{rz}z}]_{\text{tan}} = [\mathbf{A}_t e^{-jk_{ty}y} e^{-jk_{tz}z}]_{\text{tan}} \qquad (3.3)$$

This equation must be satisfied at all points on the interface, that is, for all values of y and z. We note that each specification of a point (y, z) leads to an equation in terms of the unknowns $\mathbf{A}_r$, $\mathbf{A}_t$, k_{ry}, k_{rz}, k_{ty}, and k_{tz}. Obviously, we can specify enough of such points to have more equations than unknowns leading to an inconsistant set of requirements. The only nontrivial solution is to require that

$$k_{iy} = k_{ry} = k_{ty} \equiv k_y$$
$$k_{iz} = k_{rz} = k_{tz} \equiv k_z \qquad (3.4)$$

in which case Eq. 3.3 does not depend on our choice of (y, z). These relations are known as the phase-matching requirements. Physically, they imply that the incident, reflected, and transmitted wave vectors lie in the same plane. Without loss of generality, then, let us rotate our coordinate system so that all three wave vectors lie in the xz plane as shown in Figure 3.3. This plane is called the plane of incidence and is not to be confused with the interface plane that separates regions 1 and 2. We denote the angle that the incident, reflected, and transmitted waves make with the x axis by θ_i, θ_r, and θ_t, respectively. In terms of these angles we have

$$\mathbf{k}_i = -\hat{x}k_{ix} + \hat{z}k_{iz}$$

$$\mathbf{k}_r = +\hat{x}k_{rx} + \hat{z}k_{rz}$$

$$\mathbf{k}_t = -\hat{x}k_{tx} + \hat{z}k_{tz}$$

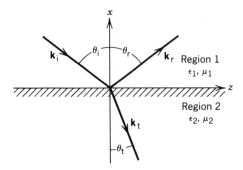

Figure 3.3 Relative orientation between incident, reflected, and transmitted wave vectors in the plane of incidence.

where

$$k_{ix} = k_1 \cos \theta_i \qquad k_{iz} = k_1 \sin \theta_i$$

$$k_{rx} = k_1 \cos \theta_r \qquad k_{rz} = k_1 \sin \theta_r$$

$$k_{tx} = k_2 \cos \theta_t \qquad k_{tz} = k_2 \sin \theta_t$$

and

$$k_1 = \omega\sqrt{\mu_1\epsilon_1} \qquad k_2 = \omega\sqrt{\mu_2\epsilon_2}$$

It is important to note that the x components of the incident and transmitted **k** vectors are negative since, as is shown in Figure 3.3, the associated plane waves travel in a direction having a negative $\hat{x}$ component. The requirement, Eq. 3.4, that the tangential or z component of the incident, reflected, and transmitted wave vectors are the same, therefore, yields from above

$$\sin \theta_i = \sin \theta_r$$

and
$$k_1 \sin \theta_i = k_2 \sin \theta_t$$

(3.5)

Thus, the angle of incidence must equal the angle of reflection and the transmitted angle is related to the angle of incidence by

$$\frac{\sin \theta_i}{\sin \theta_t} = \sqrt{\frac{\mu_2\epsilon_2}{\mu_1\epsilon_1}}$$

(3.6)

When both regions have the same permeability the latter relation is known as Snell's law.

Example 3.1

A plane wave is incident from free space, region 1, upon a region 2 having $\mu = \mu_0$ and $\epsilon = 2\epsilon_0$. The incident wave vector is $\mathbf{k}_i = -\hat{x}(2\pi/3) + \hat{z}(4\pi/3)$.

(a) What is the functional form of the spatial variation of the incident wave with respect to x and z?
(b) What is the transmitted vector, $\mathbf{k}_t$?

Solution

(a) The amplitude of our plane wave is proportional to the factor $e^{-j\mathbf{k}\cdot\mathbf{r}}$. Thus, the incident wave is proportional to

$$\exp\left\{-j\left[\left(-\hat{x}\frac{2\pi}{3} + \hat{z}\frac{4\pi}{3}\right)\cdot(\hat{x}\,x + \hat{z}\,z)\right]\right\} = e^{j(2\pi/3)x}e^{-j(4\pi/3)z}$$

Again, it is important to note that the *positive j* in the above exponent corresponds to a component of propagation along the *negative x* direction.

(b) The z component of $\mathbf{k}_t$ is already known since by our phase-matching arguments it is the same as in region 1. The x component of $\mathbf{k}_t$ can be obtained from the dispersion relation in region 2:

$$k_{tx} = \sqrt{k_2^2 - k_z^2}$$

To obtain k_2 we note that

$$k_1 = \sqrt{k_{ix}^2 + k_z^2} = \frac{2\pi}{3}\sqrt{5} = \omega\sqrt{\mu_0\epsilon_0}$$

and

$$k_2 = \omega\sqrt{\mu_0\,2\epsilon_0} = \sqrt{2}k_1 = \frac{2\pi}{3}\sqrt{10}$$

Thus

$$k_{tx} = \sqrt{\left(\frac{2\pi}{3}\right)^2(10) - \left(\frac{4\pi}{3}\right)^2} = \frac{2\pi}{3}\sqrt{6}$$

The results just derived can be envisioned in a simple graphical manner. We note that waves in regions 1 and 2 both satisfy dispersion relations of the form

$$k_x^2 + k_z^2 = k_{1,2}^2$$

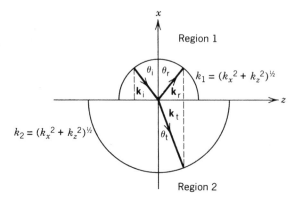

Figure 3.4 Graphical demonstration of the phase-matching concept when $k_1 < k_2$.

which represents the equation of a circle in **k** space. The magnitude of the wavenumber for the incident and reflected waves must, therefore, lie on a circle of radius k_1 while that of the transmitted wave must lie on one of radius k_2. This is shown in Figure 3.4 for the case where $k_1 < k_2$ and we have superimposed our circles in **k** space upon the geometry for the true physical boundary.

Once the k-space circles are drawn and the angle of incidence θ_i specified, phase matching requires the projections of $\mathbf{k}_r$ and $\mathbf{k}_t$ onto the z axis to be the same, thereby determining θ_r and θ_t as shown. We note that when $k_1 < k_2$ we can determine a value of θ_t for any θ_i. Let us next examine what happens when $k_1 > k_2$ as shown graphically in Figure 3.5. Because $k_1 > k_2$, we see that the transmitted angle θ_t is always greater than θ_i. However, as the incident angle approaches the so-called critical angle θ_c the transmitted angle θ_t approaches 90° as shown in Figure 3.6. For incident angles greater than θ_c we are unable to find graphically a transmitted direction because the projection of the $\mathbf{k}_t$ onto the $\hat{z}$ axis is required to exceed its magnitude. From Figure 3.6 we see

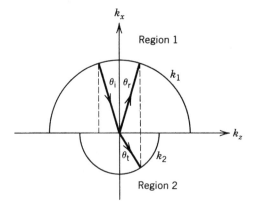

Figure 3.5 Graphical demonstration of phase matching when $k_1 > k_2$ and θ_i is less than θ_c.

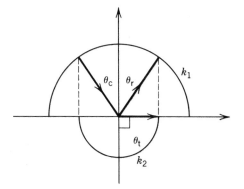

Figure 3.6 Location of the incident, reflected, and transmitted wave vectors when $\theta_i = \theta_c$.

that the θ_c is determined by the relation

$$k_1 \sin \theta_c = k_2$$

or

$$\sin \theta_c = k_2/k_1 = \sqrt{\epsilon_2/\epsilon_1} \qquad (3.7)$$

To examine the form of the transmitted **k** vector for $\theta_i > \theta_c$, we solve for the $\hat{x}$ component k_{tx} in terms of θ_i. The dispersion relation in region 2 gives

$$k_{tx} = \sqrt{k_2^2 - k_z^2}$$

and because of phase matching, $k_z = k_1 \sin \theta_i$ so that

$$k_{tx} = \sqrt{k_2^2 - k_1^2 \sin^2\theta_i}$$

We note that for $k_1 \sin \theta_i > k_2$ the argument of the square root is negative. The angle for which the argument goes to zero is from Eq. 3.7 just the critical angle. For $\theta_i > \theta_c$, k_{tx} becomes purely imaginary

$$k_{tx} = \pm j\sqrt{k_1^2 \sin^2\theta_i - k_2^2} \equiv \pm j\,\alpha_{tx}$$

We note that the assumed variation with x in region 2 is of the form

$$e^{jk_{tx}x} = e^{j(\pm j\alpha_{tx})x} = e^{\mp \alpha_{tx}x}$$

Because we require a solution that is bounded as x approaches $-\infty$, then the plus sign must be chosen. Thus,

$$k_{tx} = -j\alpha_{tx} \qquad \theta_i > \theta_c \qquad (3.8)$$

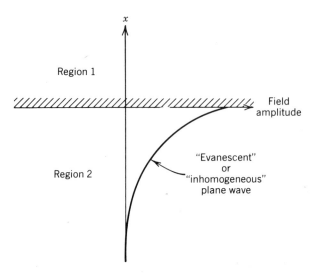

Figure 3.7 Evanescent field amplitude as a function of depth into region 2 for fixed z.

and the fields, therefore, decrease exponentially into region 2. The field in region 2 is described as being an "evanescent" or inhomogenous plane wave as shown in Figure 3.7.

3.4   REFLECTION AND TRANSMISSION COEFFICIENTS

We have obtained a significant amount of information about incident, reflected, and transmitted waves in regions 1 and 2 by an examination of the required relationships between their propagation vectors. To determine the amplitudes of the reflected and transmitted waves $\mathbf{A}_r$ and $\mathbf{A}_t$ relative to that of an arbitrarily polarized incident wave $\mathbf{A}_i$ it is convenient to decompose the incident polarization into two orthogonal components. These components are chosen to be perpendicular and parallel to the plane of incidence. If the transmitted and reflected wave amplitudes corresponding to each incident polarization are found separately, then the total fields may be obtained by superposition of the two solutions.

3.4.1   TE Wave Incidence

Let us first consider the situation when the incident wave has the electric field polarized in the plane perpendicular or transverse to the plane of incidence as shown in Figure 3.8. Note that, as required, $\mathbf{H}_i$ is perpendicular to both $\mathbf{E}_i$ and $\mathbf{k}_i$ and $\mathbf{E}_i \times \mathbf{H}_i$ is in the direction of propagation $\mathbf{k}_i$. Such a wave is said to be transverse electric or TE. Without loss of generality, we set the magnitude of the incident electric field amplitude equal to unity so that

$$\mathbf{A}_i = \hat{y}$$

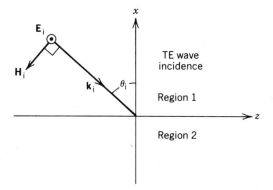

Figure 3.8 Incident field orientation for TE waves.

We assume a priori that both the reflected and transmitted waves are also polarized along $\hat{y}$ and are thus of the form

$$\mathbf{A}_r = \hat{y}R$$

$$\mathbf{A}_t = \hat{y}T$$

where R and T are complex constants. Thus, the electric fields take the form

$$\left.\begin{array}{l} \mathbf{E}_i = \hat{y}e^{+jk_{ix}x} \\[4pt] \mathbf{E}_r = \hat{y}Re^{-jk_{rx}x} \\[4pt] \mathbf{E}_t = \hat{y}Te^{+jk_{tx}x} \end{array}\right\} e^{-jk_z z}$$

The magnetic field $\mathbf{H}$ is obtainable from the curl of $\mathbf{E}$ via Maxwell's equations. For plane waves of the form $e^{-j\mathbf{k}\cdot\mathbf{r}}$, this relation has been shown to be given by

$$\mathbf{H} = \frac{1}{\omega\mu} \mathbf{k} \times \mathbf{E}$$

Thus, for example, for the incident $\mathbf{H}$ field

$$\mathbf{H}_i = \frac{1}{\omega\mu_1}(\mathbf{k}_i \times \mathbf{E}_i)$$

$$= \frac{1}{\omega\mu_1}(-\hat{x}k_{ix} + \hat{z}k_z) \times \hat{y}e^{jk_{ix}x}e^{-jk_z z}$$

$$= \frac{1}{\omega\mu_1}(-\hat{z}k_{ix} - \hat{x}k_z)e^{jk_{ix}x}e^{-jk_z z}$$

Expressions for the reflected and transmitted **H** fields are similarly obtained yielding

$$\mathbf{H}_r = \frac{R}{\omega\mu_1}(\hat{z}k_{rx} - \hat{x}k_z)e^{-jk_{rx}x}e^{-jk_z z}$$

and

$$\mathbf{H}_t = \frac{T}{\omega\mu_2}(-\hat{z}k_{tx} - \hat{x}k_z)e^{jk_{tx}x}e^{-jk_z z}$$

The requirements that tangential **E** and **H** be continuous at $x = 0$ yield, respectively, the following two relations

$$1 + R = T$$

$$1 - R = \frac{k_{tx}}{k_{ix}}\frac{\mu_1}{\mu_2}T$$

where use has been made of that fact that $k_{ix} = k_{rx}$. Solving the above expressions simultaneously for the reflection and transmission coefficients R and T yields

$$R^{\mathrm{TE}} = \frac{1 - (\mu_1/\mu_2)(k_{tx}/k_{ix})}{1 + (\mu_1/\mu_2)(k_{tx}/k_{ix})}$$

$$T^{\mathrm{TE}} = \frac{2}{1 + (\mu_1/\mu_2)(k_{tx}/k_{ix})}$$

(3.9)

where we have included the superscript to identify the coefficients as associated with TE wave incidence.

Example 3.2

Compute the reflection coefficient for a TE plane wave incident at a 30° angle from region 1 having $\mu_1 = \mu_0$ and $\epsilon_1 = 2\epsilon_0$ onto region 2, with $\mu_2 = \mu_0$ and $\epsilon_2 = \epsilon_0$.

Solution

Since the permeabilities are the same in both regions

$$R^{\mathrm{TE}} = \frac{1 - k_{tx}/k_{ix}}{1 + k_{tx}/k_{ix}}$$

Now

$$k_{ix} = k_1 \cos\theta_i$$

and

$$k_{tx} = k_2 \cos \theta_t = k_2 \sqrt{1 - \sin^2\theta_t}$$

The angle θ_t can be related to θ_i via Snell's law:

$$\sin \theta_t = \frac{k_1}{k_2} \sin \theta_i = \sqrt{\epsilon_1/\epsilon_2} \sin \theta_i$$

Thus

$$\frac{k_{tx}}{k_{ix}} = \sqrt{\epsilon_2/\epsilon_1} \; \frac{\sqrt{1 - \epsilon_1/\epsilon_2 \sin^2\theta_i}}{\cos \theta_i}$$

$$= \frac{\sqrt{\epsilon_2/\epsilon_1 - \sin^2\theta_i}}{\cos \theta_i}$$

Substituting for ϵ_1, ϵ_2, and θ_i yields

$$\frac{k_{tx}}{k_{ix}} = \frac{\sqrt{0.5 - \sin^2 30°}}{\cos 30°} = 0.577$$

and

$$R^{TE} = 0.268$$

3.4.2 TM Wave Incidence

Solutions for incident wave polarization in the plane of incidence can be obtained directly by duality. Figure 3.9 shows the relative field orientation for a wave incident with **E** in the plane of incidence. For this polarization, **H** is transverse to the plane of incidence. We note that **E** and **H** are in fact the duals of those for TE wave incidence and as a result, solutions for reflection and transmission coefficients must be the duals of those found for TE wave incidence. Thus, the reflection and transmission coefficients are found by replacing $\mu_{1,2}$ with $\epsilon_{1,2}$ yielding

$$\frac{H_r}{H_i} \equiv R^{TM} = \frac{1 - (\epsilon_1/\epsilon_2)(k_{tx}/k_{ix})}{1 + (\epsilon_1/\epsilon_2)(k_{tx}/k_{ix})}$$

$$\frac{H_t}{H_i} \equiv T^{TM} = \frac{2}{1 + (\epsilon_1/\epsilon_2)(k_{tx}/k_{ix})}$$

(3.10)

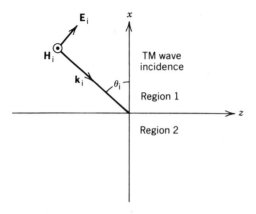

Figure 3.9 Incident field orientation for TM waves.

where the superscript indicates the results are for transverse magnetic field or TM incidence. Care should be taken to note that these dual reflection coefficients relate the ratio of *magnetic field* amplitudes and not *electric field* as was the case for TE wave incidence.

The reflection and transmission coefficients derived are in general complex, having a magnitude and phase that depend upon the angle of incidence. Let us denote the reflection coefficients as

$$R^{\text{TE}} = |R^{\text{TE}}| \, e^{j\phi^{\text{TE}}}$$

$$R^{\text{TM}} = |R^{\text{TM}}| \, e^{j\phi^{\text{TM}}}$$

Figure 3.10 shows typical plots of magnitude and phase for R^{TE} as a function of θ_i with $\epsilon_1 < \epsilon_2$ and $\epsilon_1 > \epsilon_2$, respectively. Figure 3.11 shows these same quantities for

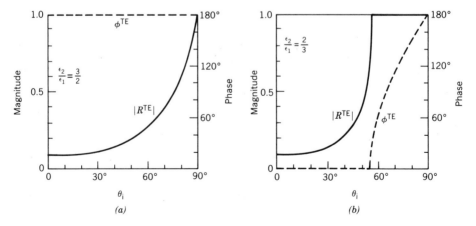

Figure 3.10 Magnitude and phase of TE reflection coefficient. (*a*) $\epsilon_2 > \epsilon_1$; (*b*) $\epsilon_2 < \epsilon_1$.

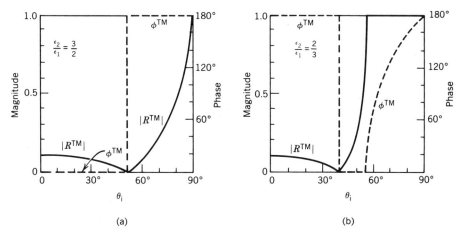

Figure 3.11 Magnitude and phase of TM reflection coefficient. (*a*) $\epsilon_2 > \epsilon_1$; (*b*) $\epsilon_2 < \epsilon_1$.

R^{TM}. In both figures, regions 1 and 2 are assumed to have free space permeabilities so that $\mu_1 = \mu_2 = \mu_0$.

Several interesting features may be observed from these figures:

1. For propagation from a less dense into a more dense medium $|R^{TE}|$ and $|R^{TM}| < 1$ for $\theta_i < \pi/2$.
2. For propagation from a less dense into a more dense medium ϕ^{TE} and ϕ^{TM} are either 0 or π.
3. For propagation from a more dense into a less dense medium $|R^{TE}| = |R^{TM}| = 1$ for $\theta_i > \theta_c$. In this region ϕ^{TE} and ϕ^{TM} vary smoothly from 0 to π.
4. For some angle $\theta_i \equiv \theta_B$, $|R^{TM}| = 0$ but for no angle θ_i is $|R^{TE}| = 0$.

Several of these properties are discussed in more detail in the following sections.

3.5 TOTAL INTERNAL REFLECTION

Let us examine in more detail the nature of reflected and transmitted fields resulting from a plane wave impinging upon a less dense medium at an angle exceeding the critical angle. We have shown in Section 3.3 that when the incident angle exceeds the critical angle, the $\hat{x}$ component of the transmitted wave vector k_{tx} becomes imaginary, that is, $k_{tx} = -j\alpha_{tx}$ where α_{tx} is real and positive and the field is evanescent or exponentially decaying in region 2. In this limit, we have

$$R^{TE} = \frac{1 + j(\alpha_{tx}/k_{ix})}{1 - j(\alpha_{tx}/k_{ix})}$$

$$R^{TM} = \frac{1 + j(\epsilon_1/\epsilon_2)(\alpha_{tx}/k_{ix})}{1 - j(\epsilon_1/\epsilon_2)(\alpha_{tx}/k_{ix})}$$

so that

$$|R^{TE}| = |R^{TM}| = 1$$

and

$$\phi^{TE} = 2 \tan^{-1}(\alpha_{tx}/k_{ix}) \qquad \theta_i > \theta_c$$

$$\phi^{TM} = 2 \tan^{-1}\left(\frac{\epsilon_1}{\epsilon_2}\frac{\alpha_{tx}}{k_{ix}}\right) \tag{3.11}$$

where

$$\frac{\alpha_{tx}}{k_{ix}} = \frac{\sqrt{\sin^2\theta_i - (\epsilon_2/\epsilon_1)}}{\cos\theta_i}$$

Thus, in agreement with Figures 3.10 and 3.11, we see that beyond the critical angle the magnitude of the reflection coefficient is unity with the phase varying continuously from 0 to π.

Further insight into the abrupt change in the nature of fields at the critical angle can be obtained by examining the behavior of incident, reflected, and transmitted power flow as θ_i is varied. The Poynting vectors are obtained from the fields given in Section 3.4.1. For TE waves, these are given, respectively, by

$$\mathbf{S}_i = \frac{1}{\omega\mu_1}(-\hat{x}k_{ix} + \hat{z}k_z)$$

$$\mathbf{S}_r = \frac{|R^{TE}|^2}{\omega\mu_1}(\hat{x}k_{ix} + \hat{z}k_z)$$

$$\mathbf{S}_t = \frac{|T^{TE}|^2}{\omega\mu_2}(-\hat{x}k_{tx}^* + \hat{z}k_z)e^{j(k_{tx}-k_{tx}^*)x}$$

with expressions for TM waves obtainable by duality. Note that for $\theta_i < \theta_c$, k_{tx} is real and, therefore, all three Poynting vectors are real and independent of x and z. In this limit, we see that $\frac{1}{2}|\mathbf{S}_i|$, $\frac{1}{2}|\mathbf{S}_r|$, and $\frac{1}{2}|\mathbf{S}_t|$ represent the time-average power flow for the incident, reflected, and transmitted waves and θ_i, θ_r, and θ_t represent the directions of the power flow as expected.

It is useful to define the ratios of the reflected and transmitted power flowing normal to the interface at $x = 0$ relative to the incident power flow. These values are known as the reflectivity and transmissivity, respectively, and are given by

$$r^{TE} = \frac{\hat{x} \cdot \mathbf{S}_r}{-\hat{x} \cdot \mathbf{S}_i}$$

$$t_{TE} = \frac{-\hat{x} \cdot \mathbf{S}_t}{-\hat{x} \cdot \mathbf{S}_i} \tag{3.12}$$

From above, we have

$$r^{TE} = |R^{TE}|^2$$
$$t^{TE} = \frac{\mu_1}{\mu_2} \frac{k_{tx}^*}{k_{ix}} |T^{TE}|^2$$

(3.13)

and by duality for TM waves

$$r^{TM} = |R^{TM}|^2$$
$$t^{TM} = \frac{\epsilon_1}{\epsilon_2} \frac{k_{tx}^*}{k_{ix}} |T^{TM}|^2$$

(3.14)

From expressions 3.9 for R^{TE} and T^{TE} and 3.10 for R^{TM} and T^{TM} it is easy to show that

$$\mathrm{Re}(r^{TE} + t^{TE}) = 1 \qquad \mathrm{Re}(r^{TM} + t^{TM}) = 1$$

(3.15)

which is a statement of conservation of time-average power.

Example 3.3

A TE plane wave is normally incident from a dense region having $\epsilon_1 = 4\epsilon_0$ onto free space. Assuming $\mu_1 = \mu_2 = \mu_0$, compute the reflectivity and transmissivity for the incident wave and verify that conservation of power holds.

Solution

For normal incidence,

$$k_{ix} = k_1 \quad \text{and} \quad k_{tx} = k_2$$

Thus

$$R^{TE} = \frac{1 - \sqrt{\epsilon_2/\epsilon_1}}{1 + \sqrt{\epsilon_2/\epsilon_1}} = \frac{1}{3}$$

and

$$T^{TE} = \frac{2}{1 + \sqrt{\epsilon_2/\epsilon_1}} = \frac{4}{3}$$

Note that

$$R^{TE} + T^{TE} = \tfrac{1}{3} + \tfrac{4}{3} = \tfrac{5}{3} \neq 1!$$

Now

$$r^{\text{TE}} = |R^{\text{TE}}|^2 = \tfrac{1}{9}$$

and

$$t^{\text{TE}} = \frac{k_{\text{tx}}^*}{k_{\text{ix}}} |T^{\text{TE}}|^2 = \sqrt{\epsilon_2/\epsilon_1} \, |T^{\text{TE}}|^2 = \tfrac{1}{2} \, (\tfrac{4}{3})^2 = \tfrac{8}{9}$$

Therefore,

$$r^{\text{TE}} + t^{\text{TE}} = \tfrac{1}{9} + \tfrac{8}{9} = 1$$

When $\theta_i > \theta_c$ $k_{\text{tx}} = -j\alpha_{\text{tx}}$ becomes imaginary. In this case, only the $\hat{z}$ component of $\mathbf{S}_t$ is real indicating that there is time-average power flowing in the $\hat{z}$ direction but not in the $\hat{x}$ direction. Further, the amount of power flowing parallel to the interface in region 2 decreases exponentially with penetration into this half space. Because the $\hat{x}$ component of $\mathbf{S}_t$ is imaginary, it represents energy stored in electric and magnetic fields that also decreases exponentially with penetration into the transmitted region.

We therefore conclude the very important result that when $\theta_i > \theta_c$ a plane wave incident upon a less dense half space is totally internally reflected with no net power flowing into the less dense medium. The electric and magnetic fields do, however, penetrate into the less dense material, decreasing exponentially away from the surface and provide a mechanism for energy storage in this region.

Let us examine explicitly the electric field distribution above and below the dielectric interface when the critical angle is exceeded. Again we consider TE incidence with the TM solution obtained by duality. To compute the total field in region 1, note that the magnitude of the reflection coefficient is unity and the phase ϕ is given by Eq. 3.11. Thus, for $x > 0$

$$E_y(x,z) = (e^{jk_{\text{ix}}x} + e^{j\phi}e^{-jk_{\text{ix}}x})e^{-jk_z z}$$

Multiplying by $e^{j\omega t}$ and taking the real part to give the true time–space variation then yields

$$E_y(x, z, t) = 2 \cos(k_{\text{ix}}x - \phi/2) \cos(\omega t - k_z z + \phi/2) \qquad (3.16)$$

To compute the transmitted field, use is made of the previously derived relation between reflection and transmission coefficient,

$$T = 1 + R$$

so that for $x < 0$

$$E_y(x, z) = (1 + e^{j\phi})e^{\alpha_{\text{tx}}x}e^{-jk_z z}$$

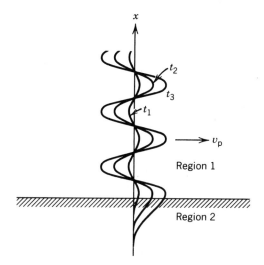

Figure 3.12 Variation in field amplitude as a function of x for a standing wave at three different times. The plot is made at some fixed but arbitrary value of z.

yielding a time-dependent variation given by

$$E_y(x, z, t) = 2 \cos(\phi/2)e^{\alpha_{tx}x}\cos(\omega t - k_z z + \phi/2) \qquad (3.17)$$

A "snapshot" of the field variation with x for fixed z at several different times would, therefore, look as shown in Figure 3.12. Note that although the amplitude of the distribution changes with time, the position of field nulls and maxima remains unchanged. Such a distribution is said to be standing wave in nature with respect to x because the distribution does not propagate along this direction. The standing wave is the direct result of zero net power flow along x.

From Eqs. 3.16 and 3.17, however, it is observed that the field distribution is not standing wave in nature with respect to z. As shown in Figure 3.12, each point along x moves to the right at a phase velocity given by $v_p = \omega/k_z$. Again this is consistent with nonzero power flow along this direction.

3.6 THE GOOS–HAENCHEN SHIFT (ADVANCED TOPIC)

It has been shown that a plane wave incident upon a less dense medium at an angle greater than the critical angle experiences not only total internal reflection but also a phase shift that is a function of the angle of incidence. While this phase shift has little meaning for a single plane wave, a physical interpretation is possible for incident beams having finite cross-sectional dimensions. As we show in this section, the effect of the angularly dependent phase shift is to displace the reflected beam parallel to the interface (along z) with respect to the incident beam.

To understand this phenomenon, first consider what happens when a well-collimated

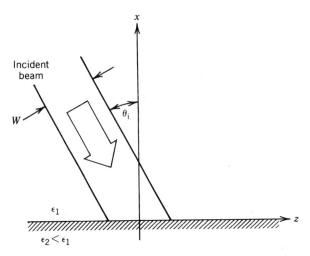

Figure 3.13 Collimated beam of width W incident from a high-permittivity region upon a lower permittivity half-space.

beam of width W is totally internally reflected at a dielectric interface as shown in Figure 3.13. For simplicity, we will assume that the beam has infinite extent in the y direction thereby making the geometry for the problem two dimensional.

When W is infinite, the incident beam is a plane wave. Assuming TE incidence at an angle θ_i, then for a unity amplitude wave,

$$\mathbf{E}_i\,(x,\,z)\,=\,\hat{y}e^{jk_{x0}x}e^{-jk_{z0}z}$$

where

$$k_{x0}\,=\,k\cos\theta_i \qquad k_{z0}\,=\,k\sin\theta_i$$

When W is finite, then the incident beam can be represented as the product of the above plane wave and the "envelope" function, $W(x,\,z)$, which truncates the plane wave for distances greater than $\pm W/2$ beyond the beam propagation axis. That is,

$$\mathbf{E}_i\,(x,\,z)\,=\,\hat{y}\,\underbrace{(e^{jk_{x0}x}e^{-jk_{z0}z})}_{\text{Constant phase fronts}}\,\times\,\underbrace{W(x,\,z)}_{\text{Beam envelope}} \qquad (3.18)$$

As indicated, the terms in parentheses determine the locations of constant phase fronts associated with the beam, whereas $W(x,\,z)$ determines where the beam is nonzero.

Now, as was shown in Chapter 2, the incident beam can also be represented by a superposition of plane waves of the form

$$\mathbf{E}_i\,(x,\,z)\,=\,\hat{y}\int_{-\infty}^{\infty}dk_x\,E(k_x)e^{jk_xx}e^{-jk_zz} \qquad (3.19)$$

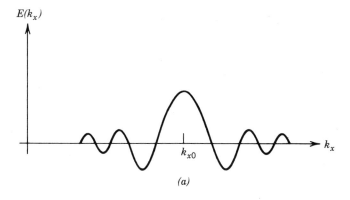

(a)

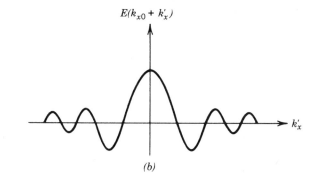

(b)

Figure 3.14 Plane-wave spectrum of the incident beam.

Further, it was shown that, provided W is many wavelengths, then $E(k_x)$ is a sharply peaked sinc function centered near $k_x = k_{x0}$ as illustrated in Figure 3.14a. Since $E(k_x)$ contributes to the above integral only near k_{x0}, let us make the substitution of variables

$$k'_x = k_x - k_{x0} \tag{3.20}$$

so that $E(k_{x0} + k'_x)$ is centered about $k'_x = 0$ as shown in Figure 3.14b. To evaluate the integral given in Eq. 3.19, it is also necessary to express k_z in terms of k'_x. From the dispersion relation we have

$$k_z = \sqrt{k^2 - k_x^2}$$

$$= \sqrt{(k^2 - k_{x0}^2) - 2 k_{x0}k'_x + (k'_x)^2}$$

$$= k_{z0} \sqrt{1 - 2 (k_{x0}/k_{z0})(k'_x/k_{z0}) + (k'_x/k_{z0})^2}$$

Since $E(k_{x0} + k'_x)$ contributes to integral 3.19 only when k'_x is very small, then the second term in the above radical is multiplied by the generally small quantity,

(k'_x/k_{x0}), and the third term multiplied by the square of this small quantity which can be neglected completely.* Therefore, using the expansion for small x,

$$(1 + x)^{1/2} \cong 1 + \tfrac{1}{2}x$$

we have

$$k_z \cong k_{z0} + k'_z \tag{3.21}$$

where

$$k'_z \equiv - \frac{k_{x0}}{k_{z0}}k'_x \tag{3.22}$$

Substituting Eqs. 3.20–3.22 into integral 3.19 then yields

$$\mathbf{E}_i(x, z) \cong \underbrace{\hat{y}e^{jk_{x0}x}e^{-jk_{z0}z}}_{\text{Phase front}} \underbrace{\int_{-\infty}^{\infty} dk'_x \, E(k_{x0} + k'_x)e^{jk'_x x}e^{-jk'_z z}}_{\text{Beam envelope}} \tag{3.23}$$

Comparing Eq. 3.23 with 3.18, we observe that the integral above is merely an alternative description for beam envelope $W(x, z)$.

To obtain the reflected beam, note that by superposition, each plane-wave component of the incident beam described in Eq. 3.23

$$\hat{y}E(k_{x0} + k'_x) \, e^{j(k_{x0}+k'_x)x} \, e^{-j(k_{z0}+k'_z)z} \tag{3.24}$$

must be reflected at the interface according to the relations developed in the previous section. These laws require that the z component of the incident and reflected **k** vectors be equal but the x components have opposite sign.

The reflected wave amplitude is obtained by multiplication of the incident wave by the reflection coefficient. Therefore, the reflected wave associated with the incident component in Eq. 3.24 above is given by

$$\hat{y}E(k_{x0} + k'_x) \, e^{-j(k_{x0}+k'_x)x}e^{-j(k_{z0}+k'_z)z} \times R^{\text{TE}}(k_{x0} + k'_x) \tag{3.25}$$

If we assume that $|R^{\text{TE}}| = 1$ over those values of k'_x for which $E(k_{x0} + k'_x)$ is appreciable (that is, all wave components making up the incident beam are totally internally reflected), then

$$R^{\text{TE}}(k_{x0} + k'_x) \cong e^{j\phi^{\text{TE}}(k_{x0}+k'_x)} \tag{3.26}$$

---

*For near normal incidence, our analysis breaks down since k_{z0} approaches zero. However, for practical integrated-optics geometries θ_i is close to $\pi/2$ and therefore our assumption is valid.

Again, since by assumption k_x' is small over the region for which $E(k_{x0} + k_x')$ is nonzero, then ϕ^{TE} can be expanded in a first-order Taylor series about k_{x0} as

$$\phi^{TE}(k_{x0} + k_x') \cong \phi^{TE}(k_{x0}) + k_x' \left.\frac{\partial \phi^{TE}}{\partial k_x'}\right|_{k_x'=0} \tag{3.27}$$

Substituting Eqs. 3.26 and 3.27 into Eq. 3.25 and integrating over k_x' yields

$$\mathbf{E}_r(x, z) = \hat{y} \underbrace{e^{-jk_{x0}x}e^{-jk_{z0}z}e^{j\phi^{TE}(k_{x0})}}_{\text{Beam phase front}} \tag{3.28}$$

$$\times \underbrace{\int_{-\infty}^{\infty} dk_x' \, E(k_{x0} + k_x') \, e^{-jk_x'(x + \Delta x)}e^{-jk_z'z}}_{\text{Beam envelope}}$$

where

$$\Delta x \equiv -\left.\frac{\partial \phi^{TE}}{\partial k_x'}\right|_{k_x'=0} \tag{3.29}$$

By comparing Eqs. 3.23 and 3.28, we observe that in the special case where $\partial \phi^{TE}/\partial k_x' = 0$, then $\Delta x = 0$ and the reflected beam is given by

$$\mathbf{E}_r(x, z) = \hat{y} \, e^{-jk_{x0}x} \, e^{-jk_{z0}z} \, e^{j\phi^{TE}(k_{x0})} \times W(-x, z)$$

In this situation, the reflected beam is identical in shape to the incident beam but travels upward and away from the interface at an angle θ_i, as shown in Figure 3.15a. The reflected beam is also phase shifted by the amount $\phi^{TE}(k_{x0})$.

In reality, however, $\partial \phi^{TE}/\partial k_x' \neq 0$ and therefore $\Delta x \neq 0$. In this case,

$$E_r(x, z) = \hat{y} \, e^{-jk_{x0}x} \, e^{-jk_{z0}z} \, e^{j\phi^{TE}(k_{x0})} \times W[-(x + \Delta x), z]$$

so that the reflected beam appears to be shifted downward with respect to x by an amount Δx. This result is shown in Figure 3.15b. Thus, the beam appears to have emanated from a point located a distance Δx below the interface. The beam also appears to be shifted laterally by an amount Δz. From the geometry it is easy to show that

$$\Delta z = \Delta x \tan \theta_i$$

Although this lateral or Goos–Haenchen shift, named for its discoverers, is generally small it plays an important role in the understanding of waveguide coupling to be studied in later chapters.

In Figure 3.15c, we show an alternative interpretation: that the beam is in fact reflected from a fictitious boundary located at $x = -d = -\Delta x/2$. By computing Δx

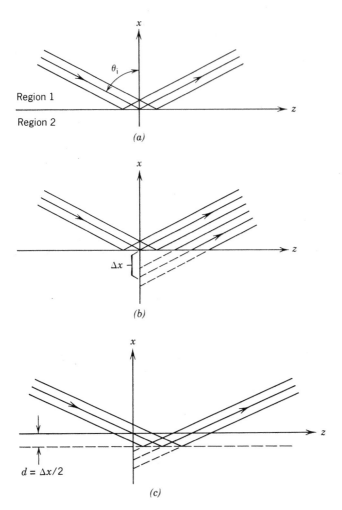

Figure 3.15 Incident and reflected beam. (*a*) No phase shift at the interface; (*b*) lateral displacement (Goos–Haenchen shift) caused by total internal reflection; (*c*) position of fictitious reflecting plane causing the equivalent lateral shift.

from Eq. 3.29, it is found that for TE and TM wave incidence, respectively, the distance below the true interface is given by

$$d^{\text{TE}} = \frac{1}{\alpha_{\text{tr}}}$$

$$d^{\text{TM}} = q\frac{1}{\alpha_{\text{tr}}}$$

(3.30)

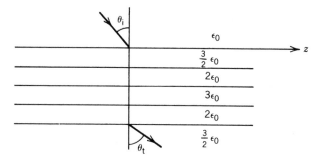

Figure 3.16

where

$$q \equiv \left(\frac{k_{ix}^2 + \alpha_{tx}^2}{k_{ix}^2 + (\epsilon_1/\epsilon_2)^2 \, \alpha_{tx}^2} \right) \frac{\epsilon_1}{\epsilon_2}$$

For geometries that are practical to integrated-optics applications, $\epsilon_1/\epsilon_2 \cong 1$ so that $q \cong 1$ and d is close in value to the $1/e$ penetration depth of the evanescent fields.

PROBLEMS

3.1 A plane wave is incident from free space upon a dielectric stack consisting of layers of different permittivity as shown in Figure 3.16:

If the angle of incidence θ_i is 25°, what is the angle for the transmitted wave θ_t?

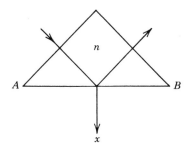

Figure 3.17

3.2   As discussed in Section 3.4.2 and shown in the example of Figure 3.11, there exists some angle of incidence θ_B for which the reflection coefficient for a TM wave goes to zero. This angle is known as the Brewster angle. Verify by setting $R^{TM} = 0$ in Eq. 3.10 that θ_B is given by the relation

$$\tan \theta_B = \sqrt{\epsilon_2/\epsilon_1}$$

You may assume $\mu_1 = \mu_2 = \mu_0$.

3.3   A beam of light from a He–Ne laser ($\lambda_0 = 0.63 \ \mu m$) is totally internally reflected by a right prism having refractive index $n = \sqrt{\epsilon/\epsilon_0} = 1.5$ as shown in Figure 3.17:

At what distance x below the prism surface AB is the wave amplitude reduced by $1/e$ of its value at the surface?

3.4   A plane wave impinges upon a dielectric interface from above with an incident wave vector $\mathbf{k}_i = k_0 \, (-\hat{x} + \sqrt{3} \, \hat{z})$, where $k_0 = \omega \sqrt{\mu_0 \epsilon_0}$.
   (a) What is the permittivity of the upper dielectric region?
   (b) What must the permittivity of the lower region be in order that the transmitted plane wave has $\theta_t = 30°$?

3.5   A TE plane wave traveling in more dense region of permittivity ϵ_1 is incident at a 45° angle upon a less dense region of permittivity ϵ_2. If the wave is totally internally reflected, what must be the ratio of ϵ_2/ϵ_1 for the phase shift upon reflection to be equal to 60°?

3.6   Make a plot of the magnitude and phase of the reflection coefficient for a TE wave incident from region 1 having $\epsilon_1 = \epsilon_0$ and $\mu_1 = \mu_0$ onto region 2 having $\epsilon_2 = 2\epsilon_0$ and $\mu_2 = \mu_0$.

3.7   Repeat 3.6 for a TM wave.

3.8   Make a plot of the magnitude and phase of the transmission coefficient for the geometry described in problem 3.6.

3.9   Repeat problem 3.8 for a TM wave.

3.10   A TM plane wave of frequency ω is incident from a medium having μ_1 and ϵ_1 onto a half space having μ_2 and $\epsilon_2 < \epsilon_1$. The amplitude of the incident magnetic field is unity and the angle of incidence is greater than the critical angle.
   (a) What are the incident electric and magnetic fields $\mathbf{E}_i(x, z)$ and $\mathbf{H}_i(x, z)$?
   (b) What are the reflected and transmitted electric and magnetic fields $\mathbf{E}_r(x, z)$, $\mathbf{E}_t(x, z)$, $\mathbf{H}_r(x, z)$, $\mathbf{H}_t(x, z)$?
   (c) Compute the incident, reflected, and transmitted Poynting vectors $\mathbf{S}_i$, $\mathbf{S}_r$, and $\mathbf{S}_t$ associated with the fields.
   (d) Verify that $\mathrm{Re}(r^{TM} + t^{TM}) = 1$.

3.11   What is the total time-varying magnetic and electric field $\mathbf{H}(x, z, t)$ and $\mathbf{E}(x, z, t)$ in each region of problem 3.10 when the incident angle is *less than* the critical angle?

REFERENCES

1. Fowles, G. R. *Introduction to Modern Optics.* New York: Holt, Rinehart, Winston, 1968.

2. Kong, J. A. *Theory of Electromagnetic Waves.* New York: Wiley, 1975.

3. Midwinter, J. E. *Optical Fibers for Transmission.* New York: Wiley, 1979.

4. Paul, C. R., and Nasar, S. A. *Introduction to Electromagnetic Fields.* New York: McGraw-Hill, 1982.

5. Tamir, T., ed. *Integrated Optics.* New York: Springer-Verlag, 1975.

CHAPTER 4

The Slab Dielectric Waveguide

4.1 INTRODUCTION

In Chapter 3, it was demonstrated that a plane wave or collimated beam incident upon a less dense medium at an angle exceeding the critical angle was totally internally reflected. In this situation the time-average Poynting vector normal to the interface was found to be zero with net power flow existing only parallel to the interface, that is, along z. The effect of the dielectric discontinuity was, therefore, to redirect or steer the incident beam away from the interface.

Suppose now that a second boundary is introduced a distance d above the first, creating a new geometry in which a high dielectric material called the core is surrounded by two lower dielectric regions. The geometry is shown in Figure 4.1.

Consider the consequences of generating in some manner within the core a collimated beam propagating, say initially, upward at an angle θ exceeding the critical angle for both upper and lower interfaces. Then, as can be observed from the geometry of Figure 4.2, the beam will continue to bounce between lower and upper boundaries at the same angle as it travels along the z axis of the plate. We show later in the chapter that the angle θ cannot be arbitrary but is constrained to a discrete set of values, with each value corresponding to a waveguide mode.

Based upon the previous analysis of total internal reflection at a single dielectric interface, the bouncing beam would be expected to have a field distribution which is standing wave in nature in the x direction within the high dielectric core and which is evanescent in the surrounding lower dielectric regions, decaying as an observer moves away from the interfaces located at $x = \pm d/2$. Further, because of the evanescent nature of these fields, no net power should propagate away from the waveguide core. Thus, we anticipate the beam to be confined in the proximity of the high dielectric region while being guided along the z direction. Such a structure is referred to as a dielectric waveguide. The simple slab waveguide described above is analyzed in detail throughout this chapter not only because of its practical importance but also because it will provide a great deal of insight into the more sophisticated guiding geometries discussed in later chapters.

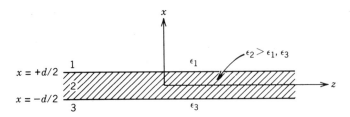

Figure 4.1 Slab dielectric waveguide.

4.2   THE SYMMETRIC SLAB DIELECTRIC WAVEGUIDE

To obtain an understanding of the behavior of guided-wave solutions we consider first, for simplicity, the symmetric slab structure. Let us, therefore, set $\epsilon_1 = \epsilon_3 = \epsilon_0$ and $\epsilon_2 = \epsilon$. Such a geometry could correspond, for example, to an idealization of a glass microscope slide surrounded by air.

Let us define a dielectric waveguide mode as a set of electromagnetic fields which maintain their transverse spatial distribution while traveling along a direction of propagation. If the propagation direction is chosen to be along z, then $\mathbf{E}$ and $\mathbf{H}$ are assumed to be of the form

$$\begin{Bmatrix} \mathbf{E}(x, y, z) \\ \mathbf{H}(x, y, z) \end{Bmatrix} = \begin{Bmatrix} \mathbf{E}(x, y) \\ \mathbf{H}(x, y) \end{Bmatrix} e^{-jk_z z}$$

where k_z is the propagation constant. For the slab dielectric waveguide, the symmetry shown in Figure 4.1 indicates that without loss of generality our coordinate system may always be rotated about the x axis in such a manner as to have all field variations lie in the xz plane and therefore independent of y.

The general form for the fields in each region, $m = 1, 2, 3$ is already known. In the last chapter it was shown that the solution to the wave equation in each region consists of TE or TM polarized plane waves that are either homogenous or inhomogenous depending on whether the transverse wavenumber in the mth region k_{mx} is purely real or purely imaginary. Based upon our heuristic explanation of guidance, we anticipate that for guided-wave solutions the electric and magnetic fields should be evanescent, that is, exponentially decaying as x approaches $\pm\infty$. In regions 1 and 3, which surround the core, the transverse wavenumber should, therefore, be imaginary. In the

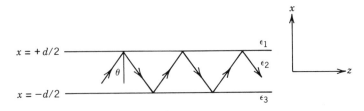

Figure 4.2 Light ray path due to total internal reflection.

core region, 2, we appeal to the anticipated standing-wave nature of the field distribution resulting from the superposition of upward and downward propagating-wave solutions, implying that the transverse wavenumber should be real. Further, due to the waveguide symmetry, it is expected that field solutions should be either even or odd with respect to x.

4.2.1   TE Solutions

Let us now attempt to find TE-mode waveguide solutions, which by definition have the electric field polarized along y. The analysis for TM modes follows along analogous lines and is presented as part of a more general analysis in Section 4.3. Based upon the above description of the anticipated field distribution we hypothesize fields of the following form:

$$E_y(x, z) = \begin{cases} A_1 e^{-\alpha_x x} & x > d/2 \\ A_2 \begin{bmatrix} \cos(k_{2x}x) \\ \sin(k_{2x}x) \end{bmatrix} e^{-jk_z z} & |x| \leq d/2 \\ \pm A_1 e^{\alpha_x x} & x < -d/2 \end{cases}$$

where the upper and lower choices for the fields correspond to even and odd solutions, respectively. The constants k_{2x} and α_x are obtained from the dispersion relations in the core and surrounding dielectric regions and are given by

$$k_{2x} = \sqrt{\omega^2 \mu \epsilon - k_z^2} \tag{4.1}$$

$$\alpha_x = \sqrt{k_z^2 - \omega^2 \mu \epsilon_0} \tag{4.2}$$

where we assume $\mu = \mu_0$ is the permeability of free space. The unknown amplitude coefficients A_1 and A_2 are to be related to each other through the requirement of continuity of tangential $\mathbf{E}$ and $\mathbf{H}$ at $x = \pm d/2$. The tangential component of $\mathbf{H}$ is obtained directly from Maxwell's curl equation,

$$H_z(x, z) = \frac{j}{\omega \mu} \frac{\partial}{\partial x} E_y(x, z)$$

yielding

$$H_z(x, z) = \frac{-j}{\omega \mu} \begin{cases} \alpha_x A_1 e^{-\alpha_x x} & x > d/2 \\ \pm k_{2x} A_2 \begin{bmatrix} \sin k_{2x}x \\ \cos k_{2x}x \end{bmatrix} e^{-jk_z z} & |x| \leq d/2 \\ \mp \alpha_x A_1 e^{+\alpha_x x} & x < -d/2 \end{cases}$$

Because of the waveguide symmetry with respect to the x axis, it is sufficient to match boundary conditions at $x = +d/2$; those at $x = -d/2$ then will be satisfied automatically. Applying these conditions at $x = +d/2$ yields, for example, for even modes

$$E_{\text{tan}} : \quad A_1 e^{-\alpha_x d/2} = A_2 \cos(k_{2x} d/2) \tag{4.3}$$

$$H_{\text{tan}} : \quad A_1 e^{-\alpha_x d/2} = \frac{k_{2x}}{\alpha_x} A_2 \sin(k_{2x} d/2) \tag{4.4}$$

Note that either expression 4.3 or 4.4 serves to relate the field amplitude in the core to that of the surrounding free-space regions, but does not uniquely determine both A_1 and A_2. If, for example, use of relation 4.3 is made we obtain for the electric fields for the even modes

$$E_y(x, z) = A_2 \begin{cases} \cos(k_{2x} d/2) e^{-\alpha_x(x-d/2)} & x > d/2 \\ \cos(k_{2x} x) & |x| \le d/2 \\ \cos(k_{2x} d/2) e^{+\alpha_x(x+d/2)} & x < -d/2 \end{cases} e^{-jk_z z} \tag{4.5}$$

Even solutions

For both relations 4.3 and 4.4 to be simultaneously true it is further required that

$$\tan(k_{2x} d/2) = \alpha_x / k_{2x} \qquad \text{(Even solutions)} \tag{4.6}$$

which is obtained directly by division of Eq. 4.4 by Eq. 4.3. Equation 4.6 is known as the guidance condition. Similar calculations for odd modes yield for the field solutions

$$E_y(x, z) = A_2 \begin{cases} \sin(k_{2x} d/2) e^{-\alpha_x(x-d/2)} \\ \sin(k_{2x} x) \\ -\sin(k_{2x} d/2) e^{+\alpha_x(x+d/2)} \end{cases} e^{-jk_z z} \tag{4.7}$$

Odd solutions

and the guidance condition

$$\cot(k_{2x} d/2) = -\alpha_x / k_{2x} \qquad \text{(Odd solutions)} \tag{4.8}$$

The transcendental relations for odd and even modes can be solved either graphically or numerically. Let us plot both sides of the guidance condition for even modes as a function of the unknown parameter $(k_{2x} d/2)$. This requires expressing α_x as a function of k_{2x}. From Eqs. 4.1 and 4.2

$$\alpha_x^2 = k_z^2 - \omega^2\mu\epsilon_0$$

$$k_{2x}^2 = \omega^2\mu\epsilon - k_z^2$$

so that we obtain

$$\alpha_x^2 = \Delta k^2 - k_{2x}^2 \tag{4.9}$$

where

$$\Delta k^2 \equiv \omega^2\mu(\epsilon - \epsilon_0)$$

The guidance condition for even modes is, therefore, given by

$$\tan(k_{2x}d/2) = \frac{\sqrt{(\Delta kd/2)^2 - (k_{2x}d/2)^2}}{(k_{2x}d/2)} \tag{4.10}$$

The left and right sides of this relation are plotted in Figure 4.3 for several values of the parameter $(\Delta kd/2)$, with the intersection points representing the allowed solutions. Note that in general more than one solution exists for a given value of $(\Delta kd/2)$. For fixed waveguide parameters ϵ and d, increasing frequency introduces more and more propagating modes. Let the pth solution be denoted by $(k_{2x})_p$. Then for any mode

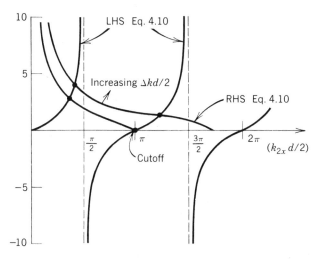

Figure 4.3 Graphical solution to the guidance condition.

solution

$$p\pi < (k_{2x})_p d/2 < (p + \tfrac{1}{2})\pi \qquad p = 0,1,2,... \qquad (4.11)$$

Further, from the dispersion relation 4.1 in region 2, once $(k_{2x})_p$ is found, the propagation constant associated with the pth mode, k_{zp}, is also uniquely determined and is given by

$$k_{zp} = \sqrt{\omega^2\mu\epsilon - (k_{2x})_p^2}$$

We now examine the mode behavior in more detail in two limits.

Limit I.   Low-Frequency Limit (Near Cutoff)

From Figure 4.3 it is observed that the solution for the pth mode is lost as

$$(k_{2x})_p d/2 \longrightarrow p\pi$$

Note additionally from Figure 4.3 that as the above condition is approached

$$\alpha_x \longrightarrow 0 \qquad \text{(Cutoff limit)} \qquad (4.12)$$

This limit is known as cutoff. As cutoff is approached, the fields in regions 1 and 3 extend further and further beyond the central guiding core. The field variation with x as this limit is approached is shown in Figure 4.4 for the first two even modes. From Figure 4.3 it is observed that as frequency is decreased fewer and fewer modes may

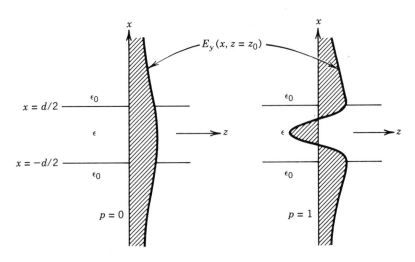

Figure 4.4 Electric field distribution for first two even TE modes near cutoff.

propagate. The frequency for which any given mode is lost is known as the cutoff frequency for that mode. Since in this limit $(k_{2x})_p d/2$ approaches $p\pi$ and α_x approaches zero, then from Eq. 4.9

$$\Delta k d/2 \longrightarrow p\pi$$

or

$$\omega\sqrt{\mu(\epsilon - \epsilon_0)} \longrightarrow 2p\pi/d$$

Thus, the cutoff frequency for the pth even mode is given by

$$f_{cp} = \frac{p}{d} \frac{c}{\sqrt{n^2 - 1}} \tag{4.13}$$

where c is the velocity of light in free space, 3×10^8 m/s, and $n = \sqrt{\epsilon/\epsilon_0}$ is the core refractive index. Note that the fundamental even mode for the symmetric slab waveguide ($p = 0$) has no cutoff frequency, and is, therefore, capable of propagating at all frequencies.

Limit II.   High-Frequency Limit (Far from Cutoff)

From Figure 4.3 it is observed that for sufficiently high frequencies

$$(k_{2x})_p d/2 \longrightarrow (p + \tfrac{1}{2})\pi \qquad \text{(High-frequency limit)} \tag{4.14}$$

Substitution of this result into field expressions 4.5 for the electric field shows that in this limit the fields are contained entirely within the dielectric core. For any finite frequency, the fields extend somewhat outside the guide but decay rapidly above and below the core. The field amplitude for the first two modes in this limit is shown in Figure 4.5.

The high- and low-frequency limits have physical significance in terms of the concept

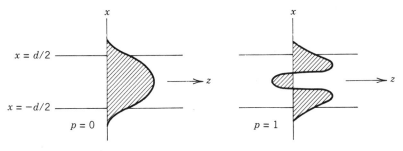

Figure 4.5 Electric field distribution for first two even TE modes far from cutoff.

of bouncing waves within the waveguide core. Note that from either expression 4.5 or 4.7 for even- and odd-mode field distributions, within the core the fields are representable as a superposition of upward and downward plane waves of the form

$$E_y(x, z) \propto (e^{+jk_{2x}x} + e^{-jk_{2x}x})e^{-jk_z z}$$

Let the propagation angle shown in Figure 4.2 associated with the pth mode be θ_p. Then the geometry indicates that

$$\tan(\theta_p) = k_{zp}/(k_{2x})_p \qquad (4.15)$$

Thus, as shown in Figure 4.6, each allowable mode propagates at a specific angle with respect to the waveguide normal and only a discrete set of angles are allowed. These angles are determined by the guidance condition which is frequency dependent.

At the high-frequency limit it has been shown that $(k_{2x})_p d/2$ approaches a constant $(p + \frac{1}{2})\pi$ whereas $k_{zp} = \sqrt{\omega^2 \mu \epsilon - (k_{2x})_p^2}$ approaches $\omega\sqrt{\mu\epsilon}$.

Therefore, from Eq. 4.15, as ω is increased, θ_p approaches $\pi/2$, indicating that the bouncing plane waves travel more nearly parallel to the waveguide walls.

As the frequency is reduced toward cutoff, α_x approaches zero whereas from Eq. 4.9, $(k_{2x})_p$ approaches Δk and from Eq. 4.2, k_z approaches $\omega\sqrt{\mu\epsilon_0}$. Therefore, in this limit,

$$\tan \theta_p = \omega\sqrt{\mu\epsilon_0}/\omega\sqrt{\mu(\epsilon - \epsilon_0)}$$

or equivalently,

$$\sin \theta_p = \sqrt{\epsilon_0/\epsilon}$$

But, from Chapter 3, this is precisely the critical angle. Therefore, the meaning of cutoff is clear; as frequency is reduced the plane-wave bounce angle within the waveguide core decreases until total internal reflection is no longer possible. In terms of

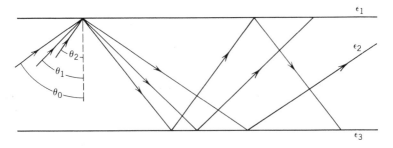

Figure 4.6 Discrete bounce angles for first three guided modes.

the bouncing-beam concept, below the cutoff frequency the mode cannot exist because a portion of the beam's power is transmitted out of the core into the surrounding space on each successive bounce. The amplitude of the mode must, therefore, diminish as the wave propagates along z. Under this situation, it is still possible to have modal solutions that maintain their transverse field distribution but decay exponentially in the direction of propagation. Such modes are called leaky modes and require for their solution a complex propagation constant $k_z = k'_z - jk''_z$. Their analysis is beyond the scope of this text.

4.2.2 Dispersion Relation (ω vs k_z)

To obtain information about phase and group velocity, it is necessary to examine the relationship between ω and k_z. This is accomplished by noting that for a given ω, the solution of the guidance condition yields a normalized transverse wavenumber $(k_{2x})_p d/2$, one for each mode. The propagation constant k_{zp} is related to $(k_{2x})_p$ and hence ω through the dispersion relation in the core given by expression 4.1. Note that each mode obeys a different ω vs k_z relationship. While a general plot of ω vs k_z for each mode must be obtained by either graphical or numerical solution, two asymptotic limits are easily examined.

Limit I.   Lower Frequency Limit

The dispersion relation in regions 1 and 3,

$$-\alpha_x^2 + k_z^2 = \omega^2 \mu \epsilon_0$$

shows that as cutoff is approached ($\alpha_x \to 0$)

$$k_z \longrightarrow \omega \sqrt{\mu \epsilon_0}$$

Thus, in this limit the propagation constant for all modes is the same and the phase velocity approaches that of free space. Simple physical arguments show that this must be the case. As cutoff is approached, the fields extend further and further outside the central core region; therefore, a larger and larger fraction of the total model power is contained in free space ($\epsilon = \epsilon_0$). Thus, in some sense we expect the mode to "see" an effective dielectric constant approaching ϵ_0.

Limit II.   High-Frequency Limit

For sufficiently high frequency we have shown that the transverse wavenumber of the pth mode approaches the limiting constant $(p + \frac{1}{2})\pi$ and therefore k_{zp} approaches $\omega \sqrt{\mu \epsilon}$. In the high-frequency limit all modes must have a phase velocity approaching

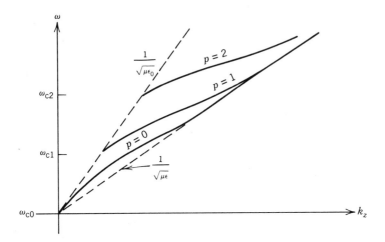

Figure 4.7 Dispersion relation for first three guided modes.

that of a plane wave propagating in an infinite medium of permittivity ϵ. Again from a physical standpoint, at high frequency the fields have been shown to be localized within the waveguide core with most of the associated modal power, therefore, confined in a region of permittivity ϵ. These results are summarized in Figure 4.7 which indicates the complete ω vs k_z diagram for a symmetric slab waveguide for several modes. Note the low- and high-frequency limits as indicated on the figure by straight lines of slope $1/\sqrt{\mu\epsilon_0}$ and $1/\sqrt{\mu\epsilon}$, respectively, as well as the cutoff frequency associated with each mode. Note that, in general, phase and group velocities are not the same, implying that the guide is dispersive.

4.3 THE ASYMMETRIC SLAB WAVEGUIDE

Although the symmetric waveguide structure is particularly easy to analyze, the asymmetric geometry is the more useful one. In practice, such a guiding structure generally consists of a substrate (region 3) which may be considered essentially infinite in thickness, with a thin guiding layer of higher refractive index above (region 2). The low refractive index, region 1, above the guiding layer is generally air, as shown in Figure 4.8. The guiding structure may be produced in a number of ways. It may, for example, consist of a glass substrate upon which a thin, slightly higher refractive glass layer has been sputtered. Alternatively, a thin, higher refractive index layer may be obtained by thermal diffusion of a metal such as titanium into a substrate such as LiNbO$_3$ or thermal diffusion of n- or p-type dopants into a semiconducting substrate such as GaAs. Typical dimensions for film thickness are on the order of 1 μm with the refractive-index differences between the film and substrate ranging between 0.1% and 10%.

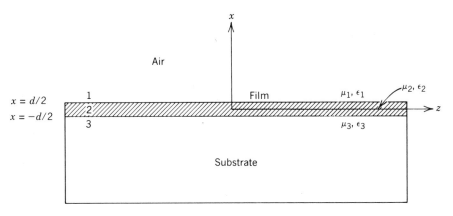

Figure 4.8 Geometry for a practical slab waveguide.

4.3.1 TE Modes

The form of the field distribution for TE modes on the asymmetric waveguide is readily extendible from those obtained for the symmetric guide. Again we assume exponentially decaying fields in regions 1 and 3 and oscillatory behavior in the core region 2. However, because the guide is no longer symmetric, the solutions will be neither even nor odd. A direct extension of the results for the symmetric guide leads us, therefore, to assume fields of the form

$$
E_y(x, z) = \begin{cases} A_1 e^{-\alpha_{1x}x} & x > d/2 \\ A_2 \cos(k_{2x}x + \psi) \\ A_3 e^{+\alpha_{3x}x} & x < -d/2 \end{cases} e^{-jk_z z} \qquad |x| \leq d/2 \qquad (4.16)
$$

where the transverse wavenumbers are defined by the appropriate dispersion relation in each region

$$
\alpha_{1x} = \sqrt{k_z^2 - \omega^2 \mu_1 \epsilon_1} \qquad (4.17)
$$

$$
\alpha_{3x} = \sqrt{k_z^2 - \omega^2 \mu_3 \epsilon_3} \qquad (4.18)
$$

$$
k_{2x} = \sqrt{\omega^2 \mu_2 \epsilon_2 - k_z^2} \qquad (4.19)
$$

Note that additionally a different permeability has been assumed in each region. While in most cases of interest all three regions are magnetically equivalent, introduction of this mathematical artifice will allow TM-mode solutions to be obtained by duality in much the same manner as was used for the dielectric half space. The constant ψ is present in the field description for region 2 because solutions are, in general, neither even nor odd. Its relationship to the amplitude coefficients A_1–A_3 is to be determined from the requirement of continuity of tangential **E** and **H** at $x = \pm d/2$. The tangential component of **H** is obtained directly from Maxwell's curl equation,

$$H_z(x, z) = \frac{j}{\omega\mu_m} \frac{\partial}{\partial x} E_y(x, z)$$

yielding

$$H_z(x, z) = \begin{cases} \dfrac{-j\alpha_{1x}}{\omega\mu_1} A_1 e^{-\alpha_{1x}x} & x > d/2 \\[3mm] \dfrac{-jk_{2x}}{\omega\mu_2} A_2 \sin(k_{2x}x + \psi) \\[3mm] \dfrac{+j\alpha_{3x}}{\omega\mu_3} A_3 e^{+\alpha_{3x}x} & x < -d/2 \end{cases} e^{-jk_z z} \quad \begin{matrix} x > d/2 \\[3mm] |x| \le d/2 \\[3mm] x < -d/2 \end{matrix}$$

Applying boundary conditions at $x = d/2$ yields

$$E_{\tan}: \quad A_1 e^{-\alpha_{1x}d/2} = A_2 \cos(k_{2x}d/2 + \psi) \tag{4.20}$$

$$H_{\tan}: \quad A_1 e^{-\alpha_{1x}d/2} = \frac{\mu_1 k_{2x}}{\mu_2 \alpha_{1x}} A_2 \sin(k_{2x}d/2 + \psi) \tag{4.21}$$

Taking the ratio of these two equations to eliminate A_1 and A_2 gives

$$\tan(k_{2x}d/2 + \psi) = \frac{\mu_2 \alpha_{1x}}{\mu_1 k_{2x}} \tag{4.22}$$

Similarly, matching of boundary conditions at $x = -d/2$ yields

$$\tan(k_{2x}d/2 - \psi) = \frac{\mu_2 \alpha_{3x}}{\mu_3 k_{2x}} \tag{4.23}$$

Noting that

$$\tan x = \tan(x \pm n\pi)$$

Eqs. 4.22 and 4.23 can be written as

$$k_{2x}d/2 + \psi = \tfrac{1}{2}\phi_1^{TE} \pm n\pi \tag{4.24}$$

$$k_{2x}d/2 - \psi = \tfrac{1}{2}\phi_3^{TE} \pm m\pi \tag{4.25}$$

where

$$\phi_1^{TE} = 2 \tan^{-1}(\mu_2 \alpha_{1x}/\mu_1 k_{2x}) \tag{4.26}$$

$$\phi_3^{TE} = 2 \tan^{-1}(\mu_2 \alpha_{3x}/\mu_3 k_{2x}) \tag{4.27}$$

Adding Eqs. 4.24 and 4.25 to eliminate ψ therefore yields the relation

$$2k_{2x}d - \phi_1^{TE} - \phi_3^{TE} = 2p\pi \qquad p = 0,1,\ldots \qquad (4.28)$$

Relation 4.28 is a generalization of the guidance condition previously derived for the symmetric waveguide. Its solution is obtained by expressing all quantities in terms of either k_{2x} or k_z and then solving either graphically or numerically. In general, there is more than one solution, corresponding to different values of p and representing different waveguide modes. For the special case of the symmetric waveguide, Eq. 4.28 yields the guidance condition for both even- and odd-mode solutions (see Problems).

Once the guidance condition has been solved for a particular mode, the corresponding fields in all three regions can be obtained by making use of boundary condition 4.20 to relate amplitude coefficients A_1 and A_2 at $x = +d/2$ and an analogous boundary condition at $x = -d/2$ to relate A_2 and A_3. The resulting fields are given by

$$E_y(x, z) = A_2 \begin{cases} \cos(k_{2x}d/2 + \psi)e^{-\alpha_{1x}(x-d/2)} & x > d/2 \\ \cos(k_{2x}x + \psi) & |x| \le d/2 \\ \cos(k_{2x}d/2 - \psi)e^{+\alpha_{3x}(x+d/2)} & x < -d/2 \end{cases} e^{-jk_z z} \qquad (4.29)$$

where ψ is given by either Eq. 4.24 or 4.25.

4.3.2 TM Modes

The solutions for TM modes are obtained directly by making use of duality. From Eqs. 4.26–4.29, the dual expressions become

$$2k_{2x}d - \phi_1^{TM} - \phi_3^{TM} = 2p\pi \qquad (4.30)$$

where

$$\phi_1^{TM} = 2 \tan^{-1}(\epsilon_2\alpha_{1x}/\epsilon_1 k_{2x}) \qquad (4.31)$$

$$\phi_3^{TM} = 2 \tan^{-1}(\epsilon_2\alpha_{3x}/\epsilon_3 k_{2x}) \qquad (4.32)$$

and

$$H_y(x, z) = A_2 \begin{cases} \cos(k_{2x}d/2 + \psi')e^{-\alpha_{1x}(x-d/2)} & x > d/2 \\ \cos(k_{2x}x + \psi') & |x| \le d/2 \\ \cos(k_{2x}d/2 - \psi')e^{+\alpha_{3x}(x+d/2)} & x < -d/2 \end{cases} e^{-jk_z z} \qquad (4.33)$$

where ψ' is given by the dual of either expression 4.24 or 4.25.

4.3.3 Ray Interpretation of the Guidance Condition

The general guidance condition for TE waves given by Eq. 4.28 and its dual, Eq. 4.30, can be given a simple physical description in terms of the bouncing-beam concept discussed at the beginning of the chapter. As was the case for the symmetric waveguide, Eq. 4.29 indicates that the fields within the core region can be described in terms of a superposition of an upward and downward propagating plane wave. In terms of the ray concept, each traveling wave can be thought of as a superposition of an infinite number of pencil beams, a few of which are shown in Figure 4.9. Several of these rays have been highlighted to emphasize how they may be viewed in terms of the concept of a bouncing beam. Also shown is an arbitrary constant phase plane for the downward propagating plane wave indicated by the dashed line.

Consider now two points A and C located along the path of the bouncing beam. Since A and C also are along the same constant phase plane, then the phase shift ϕ_{ABC} incurred in traveling along the zigzag path ABC must be an integral multiple of 2π. To compute ϕ_{ABC}, we note that this phase shift consists of a component due to propagation along the length of path connecting points A and C, ϕ_{path}, plus phase shifts ϕ_3^{TE} and ϕ_1^{TE} resulting from total internal reflection at the lower and upper interfaces, respectively. The magnitude of the phase shift along the path l_{ABC} is equal to the product of the wavenumber along this path, $k_2 = \omega\sqrt{\mu_0\epsilon_2}$, and the path length. Further, the sign of the phase shift is negative since by our convention a wave propagating forward along an arbitrary path z has a spatial dependence equal to e^{-jkz}, indicating that the wave accumulates negative phase with increasing positive distance.

The path length l_{ABC} is readily determined by unfolding it to give the equivalent distance $l_{ABC'}$, as shown in Figure 4.10. We observe that from the indicated geometry

$$l_{ABC} = l_{ABC'} = l_{CC'} \cos\theta = 2d\cos\theta$$

Thus,

$$\phi_{\text{path}} = -k_2 (2d\cos\theta)$$

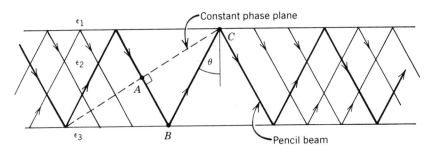

Figure 4.9 Paths for several of the "pencil beams" that constitute a guided wave. The dashed line indicates an arbitrary constant phase plane.

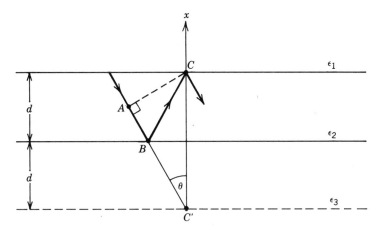

Figure 4.10 Graphical construction for computing the path length from point A to point C.

But since $k_2 \cos \theta = k_{2x}$, then

$$\phi_{\text{path}} = -2k_{2x}d$$

· Therefore we obtain the constraint for the total phase shift between A and C

$$\phi_{ABC} = \phi_{\text{path}} + \phi_1^{\text{TE}} + \phi_3^{\text{TE}} = \pm 2p\pi \qquad p = 0, 1, 2, \ldots$$

or

$$-2k_{2x}d + \phi_1^{\text{TE}} + \phi_3^{\text{TE}} = 2p\pi \qquad p = 0, 1, 2, \ldots$$

Note that negative values of p are omitted, for which no solutions exist. This represents precisely the same guidance condition given by Eq. 4.28 and by its dual, Eq. 4.30.

The constraint placed upon points A and C can also be useful for visualizing the dependence of bounce angle θ for various waveguide modes as a function of frequency. Let us suppose that the zigzag path taken by a particular mode at some arbitrary frequency ω_0 is represented in Figure 4.9. If ω is decreased, then the total phase shift ϕ_{ABC} will tend to decrease since $\phi_{\text{path}} = \omega\sqrt{\mu_0\epsilon_2}\, l_{ABC}$ is proportional to ω (we omit here the effect of ω on ϕ_1 and ϕ_3, treating this as a secondary influence). Thus, to maintain the same integral multiple of 2π phase shifts between points A and C as at ω_0, l_{ABC} must increase. This is accomplished by decreasing θ. As frequency is decreased, the angle θ continues to decrease until cutoff is reached. Note further that higher order modes have a greater number of integral multiples of 2π phase shift between points A and C. For *fixed* ω this implies that higher order modes must travel a longer path length l_{ABC}. This is accomplished by making θ successively smaller with increasing mode number. For θ sufficiently small cutoff is reached thereby limiting the number of higher order modes that may propagate.

4.3.4 Dispersion Relation for the Asymmetric Slab Waveguide

To obtain a plot of ω vs k_z for each mode, the guidance condition must be solved numerically for k_z at a number of frequencies. To make the results of the numerical computation applicable to a large number of waveguide geometries we combine the waveguide parameters into several normalized variables. A frequency parameter v which is normalized to film thickness, is defined by

$$v = k_0 \, d\sqrt{(\epsilon_2 - \epsilon_3)/\epsilon_0}$$

where

$$k_0 \equiv \omega\sqrt{\mu\epsilon_0}$$

Note that v is proportional to d/λ_0, that is, to d as measured in free-space wavelengths and also to the difference in permittivity between the core and substrate region.

Next a parameter, b is defined by

$$b = (\epsilon_{\text{eff}} - \epsilon_3)/(\epsilon_2 - \epsilon_3)$$

where the effective dielectric constant for the waveguide ϵ_{eff} is defined by

$$\frac{\omega}{k_z} = \frac{1}{\sqrt{\mu\epsilon_{\text{eff}}}}$$

To be consistent with generally accepted nomenclature, we also define the effective refractive index:

$$n_{\text{eff}} = \sqrt{\epsilon_{\text{eff}}/\epsilon_0}$$

Note that knowledge of ϵ_{eff} for each mode as a function of frequency gives the needed information on dispersion. Also, because the film permittivity ϵ_2 is generally much closer in value to that of the substrate ϵ_3 than to ϵ_1, cutoff occurs when ϵ_{eff} approaches ϵ_3 or b approaches 0. Finally, we introduce an asymmetry measure a for the TE modes on the waveguide defined by

$$a^{\text{TE}} = (\epsilon_3 - \epsilon_1)/(\epsilon_2 - \epsilon_3)$$

which varies from 0 for the symmetric waveguide to ∞ in the limit of very strong asymmetry. With these normalizations, guidance condition 4.28 for TE modes becomes

$$v\sqrt{1 - b} = p\pi + \tan^{-1}\sqrt{b/(1 - b)} + \tan^{-1}\sqrt{(b + a)/(1 - b)} \quad (4.34)$$

A numerical evaluation of the above expression yields a normalized ω vs k_z diagram as shown in Figure 4.11a for the first three modes, $p = 0, 1, 2$, and for four different asymmetry measures. Figure 4.11b shows the curves for the fundamental mode on an expanded scale and with additional a values.

For TM modes, it is found that provided the difference in refractive index between

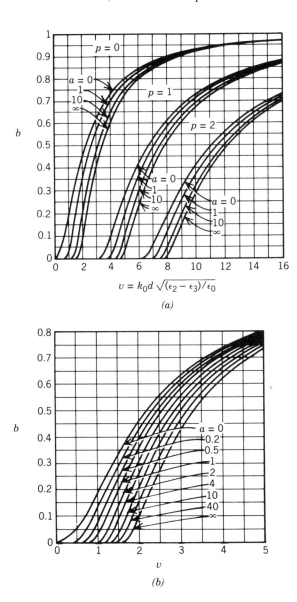

$$v = k_0 d \sqrt{(\epsilon_2 - \epsilon_3)/\epsilon_0}$$

(a)

v

(b)

Figure 4.11 Normalized b–v diagram. (a) First three modes for several values of the asymmetry measure a; (b) expanded plot for the fundamental mode. (After Kogelnik *et al.* (Ref. 5). Reproduced by permission of the authors and the Optical Society of America.)

film and substrate is small (the usual case) then the same normalized $\omega - k_z$ diagram may also be used provided that the asymmetry measure is defined by

$$a^{\text{TM}} = \left(\frac{\epsilon_2}{\epsilon_1}\right)^2 \frac{\epsilon_3 - \epsilon_1}{\epsilon_2 - \epsilon_3}$$

4.3.5   Guided-Mode Poynting Power

Let us compute the time-average Poynting power P carried by a guided mode along the direction of propagation z. To make P bounded we assume that the waveguide dimension along y is large but finite and equal to W. Then

$$P = \tfrac{1}{2} \operatorname{Re} \int_{-W/2}^{W/2} dy \int_{-\infty}^{\infty} dx (\mathbf{E} \times \mathbf{H}^*) \cdot \hat{z}$$

Since the fields are assumed to be uniform with respect to y, the integration over this variable merely results in a multiplication by W.

To facilitate the analysis, let us compute P for the even TE modes on the symmetric slab waveguide. The electric fields are given by Eq. 4.5 and the magnetic field $\mathbf{H}$ is obtained from the curl of $\mathbf{E}$. That is,

$$\mathbf{H} = \frac{j}{\omega\mu} (\hat{x} \, \partial/\partial x + \hat{z} \, \partial/\partial z) \times \hat{y} E_y$$

$$= \hat{x} H_x + \hat{z} H_z$$

Thus P is of the form

$$P = -\frac{W}{2} \operatorname{Re} \int_{-\infty}^{\infty} dx \, E_y H_x^*$$

where

$$H_x = -\frac{j}{\omega\mu} \frac{\partial E_y}{\partial z}$$

Since E_y is proportional to $e^{-jk_z z}$ then

$$H_x = -\frac{k_z}{\omega\mu} E_y$$

and

$$P = \frac{W k_z}{2\omega\mu} \int_{-\infty}^{\infty} dx \, |E_y|^2$$

Because of symmetry, the integral over x from $-\infty$ to $+\infty$ is equal to twice that between 0 and $+\infty$. This interval can in turn be divided into a portion over the high dielectric core region, $0 < x \le d/2$, plus a portion over the surrounding region, $d/2 < x$. Substitution of the appropriate field variation for $E_y(x)$ in each region yields

$$P = \frac{Wk_z}{\omega\mu}(I_1 + I_2)$$

where

$$
\begin{aligned}
I_1 &= |A|^2 \int_0^{d/2} dx\, \cos^2(k_{2x}x) \\
&= \frac{|A|^2}{2}\left(\frac{d}{2} + \frac{\sin(k_{2x}d/2)\,\cos(k_{2x}d/2)}{k_{2x}}\right)
\end{aligned}
\tag{4.35}
$$

and

$$I_2 = \frac{|A|^2}{2}\frac{\cos^2(k_{2x}d/2)}{\alpha_x}$$

The sum $I_1 + I_2$ can be put in a simpler form by noting that the guidance condition for the even TE modes can be written as

$$\cos(k_{2x}d/2) = \frac{k_{2x}}{\alpha_x}\sin(k_{2x}d/2) \tag{4.36}$$

Substituting Eq. 4.36 into Eq. 4.35 yields

$$P = \frac{Wk_z}{4\omega\mu}|A|^2 d_{\text{eff}} \tag{4.37}$$

where the effective waveguide thickness d_{eff} is defined as

$$d_{\text{eff}} = d + \frac{2}{\alpha_x}$$

Thus the amplitude coefficient A for any waveguide mode is related to the power carried by that mode via Eq. 4.37.

The expression for d_{eff} has a simple physical interpretation. Recall from the analysis of the Goos–Haenchen shift in Chapter 3 that a ray that is totally internally reflected at a dielectric interface appears to be laterally shifted. This lateral shift could be considered the result of the ray being reflected not from the true dielectric boundary but rather from a fictitious interface located slightly beyond. For TE modes, the displacement of the fictitious boundary from the true one was found to be equal to $1/\alpha_x$ where α_x is the decay constant in the less dense medium. Therefore, the interpretation of d_{eff} is clear: it represents the effective waveguide width as measured between the fictitious boundaries above and below the waveguide core that are associated with the Goos–Haenchen shift. This relationship is illustrated in Figure 4.12.

An analogous procedure may be used to relate the total power P carried by modes on the asymmetric waveguide. We find for TE and TM modes, respectively, that

$$P^{\text{TE}} = \frac{Wk_z}{4\omega\mu} |A_2|^2 d_{\text{eff}}^{\text{TE}}$$

and

(4.38)

$$P^{\text{TM}} = \frac{Wk_z}{4\omega\epsilon_2} |A_2|^2 d_{\text{eff}}^{\text{TM}}$$

Here A_2 represents the amplitude of the electric (magnetic) field in region 2 for TE (TM) waves and $d_{\text{eff}}^{\text{TE}}$ ($d_{\text{eff}}^{\text{TM}}$) is the separation between the fictitious Goos–Haenchen planes located above and below the true waveguide boundaries. Thus

$$d_{\text{eff}}^{\text{TE}} = d + \frac{1}{\alpha_{1x}} + \frac{1}{\alpha_{3x}}$$

and

(4.39)

$$d_{\text{eff}}^{\text{TM}} = d + \frac{q_1}{\alpha_{1x}} + \frac{q_3}{\alpha_{3x}}$$

where, as derived in Eq. 3.30 of Chapter 3, $q_{1,3}$ are given by

$$q_1 = \left[\frac{k_{2x}^2 + \alpha_{1x}^2}{k_{2x}^2 + \left(\dfrac{\epsilon_2}{\epsilon_1}\right)^2 \alpha_{1x}^2} \right] \frac{\epsilon_2}{\epsilon_1}$$

$$q_3 = \left[\frac{k_{2x}^2 + \alpha_{3x}^2}{k_{2x}^2 + \left(\dfrac{\epsilon_2}{\epsilon_3}\right)^2 \alpha_{3x}^2} \right] \frac{\epsilon_2}{\epsilon_3}$$

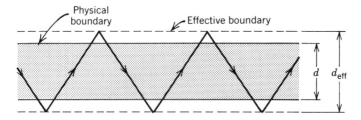

Figure 4.12 Relative locations of the true physical waveguide boundary and ''effective'' boundary resulting from Goos–Haenchen shift.

4.4 RADIATION MODES

So far we have been concerned with only those solutions for the slab dielectric waveguide having fields that decay exponentially away from the waveguide core. One may question whether other solutions exist, solutions having field distributions that are not localized near the core yet still satisfy the requirements of a mode; recall that by definition, a mode must exhibit a transverse field distribution whose magnitude is independent of position along the direction of propagation z. These solutions, as we show, do exist and are called radiation modes. They play an important role in the description of how guided radiation is coupled into and out of the core region.

4.4.1 Radiation Modes on the Symmetric Slab Waveguide

For conceptual clarity we first consider the radiation modes that may exist on the symmetric slab waveguide. These solutions will be obtained as a limiting case of the more general type of waveguide shown in Figure 4.13. The structure consists of a dielectric slab surrounded by free space with perfectly conducting plates or mirrors situated a distance $l/2$ above and below the slab. From Figure 4.13 it is observed that the slab waveguide geometry is obtained in the limit that l is allowed to approach infinity. For finite values of l it is easy to see that additional propagating modes should exist. In terms of the ray model, we can envision solutions in which the pencil beams bounce back and forth between the metallic plates or mirrors, traveling through the dielectric core at an angle that does not exceed the critical angle, as shown in Figure 4.14. Because of the waveguide symmetry, modes can again be divided into both even and odd TE and TM solutions. Let us consider the solutions for TE modes. From the ray description for the fields, solutions are anticipated that are sinusoidal rather than exponential with respect to x both in the core and the surrounding free-space regions.

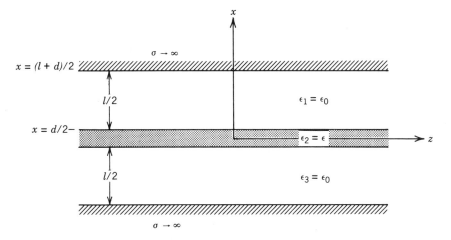

Figure 4.13 Geometry for analysis of radiation modes for the symmetric slab waveguide.

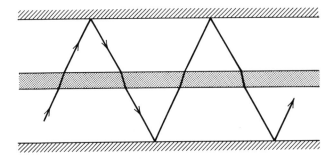

Figure 4.14 Ray path for a typical radiation mode.

We therefore assume **E** to be of the form

$$E_y(x, z) = \begin{cases} A_1 \sin[k_{1x}(x - d/2) - \psi_1] \\ A_2 \cos(k_{2x}x - \psi) \\ A_3 \sin[k_{1x}(x + d/2) + \psi_1] \end{cases} e^{-jk_z z} \qquad \begin{array}{l} d/2 < x < (d + l)/2 \\ |x| \le d/2 \\ -(d + l)/2 < x < -d/2 \end{array}$$

(4.40)

with

$$\psi_1 = k_{1x}l/2$$

Note that this choice of ψ_1 makes the electric field go to zero at the perfectly conducting plates located at $x = \pm(d + l)/2$. For even modes we require $\psi = 0$ and $A_3 = A_1$ while for odd modes $\psi = \pi/2$ and $A_3 = -A_1$.

To relate the unknown amplitude coefficient A_1 to A_2, we compute tangential **H** from **E** in the same manner as for guided modes and require continuity of both tangential **E** and **H** at $x = d/2$. This yields

$$E_{\text{tan}}: \qquad -A_1 \sin \psi_1 = A_2 \cos(k_{2x}d/2 - \psi) \qquad (4.41)$$

$$H_{\text{tan}}: \qquad A_1 \cos \psi_1 = -A_2 \frac{k_{2x}}{k_{1x}} \sin(k_{2x}d/2 - \psi) \qquad (4.42)$$

Division of relation 4.42 by Eq. 4.41 then gives the guidance condition for even and odd modes

$$\cot \psi_1 = \frac{k_{2x}}{k_{1x}} \tan(k_{2x}d/2 - \psi) \qquad \psi = \begin{cases} 0 \\ \pi/2 \end{cases} \qquad (4.43)$$

Let us now examine the behavior of the waveguide-mode solutions as l is allowed to become very large. As indicated previously, in the limit that $l = \infty$ we produce the slab dielectric waveguide geometry. To analyze this situation, guidance condition 4.43 is rewritten in terms of a single variable, k_{1x}, with k_{2x} expressed in terms of this quantity.

The relation between k_{2x} and k_{1x} is readily obtained by eliminating k_z between the dispersion relations for regions 1 and 2. We obtain straightforwardly

$$k_{2x}^2 = k_{1x}^2 + \Delta k^2$$

where

$$\Delta k^2 = \omega^2 \mu (\epsilon - \epsilon_0)$$

Thus, the guidance condition for even modes becomes

$$x \cot(xl/d) = \sqrt{x^2 + v^2} \tan\sqrt{x^2 + v^2} \qquad (4.44)$$

where

$$x = (k_{1x}d/2) \qquad v = (\Delta k\, d/2)$$

The solutions to this equation can be obtained graphically in a manner similar to that used in solving for the guided modes of the dielectric slab structure. Figures 4.15a and b show plots of the left and right sides of Eq. 4.44 for $l/d = 1$ and $l/d = 5$ with $v = 1$. The intersection points, therefore, determine a discrete set of p solutions for the transverse wavenumber in the air region $(k_{1x})_p$ and corresponding transverse wavenumber in the dielectric slab

$$(k_{2x})_p = \sqrt{\Delta k^2 + (k_{1x})_p^2} \qquad (4.45)$$

Further, each allowed value for $(k_{1x})_p$ determines a propagation constant given by

$$(k_z)_p = \sqrt{\omega^2 \mu\, \epsilon_0 - (k_{1x})_p^2}$$

A plot of the dispersion relation for a number of modes with $l/d = 1$ and $l/d = 5$ is shown in Figure 4.16. It is observed from both the graphical solution to the guidance condition as well as from the $\omega - k_z$ diagrams that the spacing of both transverse and longitudinal wavenumbers for adjacent modes decreases with increasing values of l/d. This is easily understood on physical grounds by an examination of the spatial variation of the electric fields in the x-direction. For any fixed frequency, the periodicity along x in the air region λ_{1x} and in the dielectric region λ_{2x} are determined from the transverse wavenumbers

$$2\pi/\lambda_{1x} = k_{1x}$$

$$2\pi/\lambda_{2x} = k_{2x}$$

Because k_{1x} and k_{2x} are related through phase-matching requirements, as given by Eq. 4.45, the periodicity in the air and dielectric regions are also interdependent and

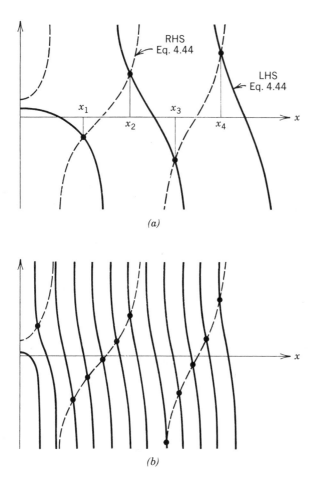

(a)

(b)

Figure 4.15 Graphical solution of the "guidance condition," Eq. 4.44. (a) $l/d = 1$, $v = 1$; (b) $l/d = 5$, $v = 1$.

specifying λ_{1x}, therefore, also determines λ_{2x}. The choice of λ_{1x} and consequently λ_{2x} is governed by the boundary conditions. These require the electric field to go to zero on the plate walls and to be continuous across the dielectric boundaries. Therefore, with reference to Figure 4.17 only those values of λ_{1x} and corresponding λ_{2x} can be chosen which smoothly fit an integral number of half "cycles" between the conductive plates. Further, as can be seen from Figure 4.17, adjacent solutions are observed to differ by one half cycle. The reason the spacing between the wavenumbers for adjacent modes decreases with increasing l now becomes clear. Since the total number of cycles of variation in the air region is proportional to the product of k_{1x} and l, then the change in k_{1x} required to add one extra half cycle between plates must decrease as l is increased. In the limit of infinite plate separation, the transverse wavenumber k_{1x} merges into a continuum. The quantities k_{2x} and k_z must also form a continuum and are related to k_{1x}

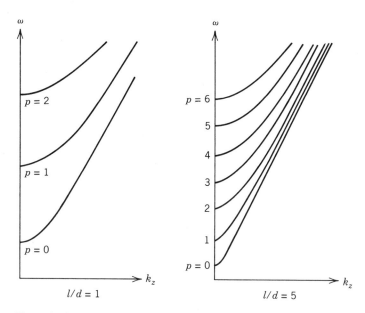

Figure 4.16 Dispersion curves. (a) $l/d = 1$, $v = 1$; (b) $l/d = 5$, $v = 1$.

via the expressions

$$k_{2x} = \sqrt{\Delta k^2 + k_{1x}^2}$$

$$k_z = \sqrt{\omega^2 \mu \, \epsilon_0 - k_{1x}^2}$$

Note that in this limit two modal solutions exist for each wavenumber, one even and one odd, corresponding to $\psi = 0$ and $\psi = \pi/2$, respectively.

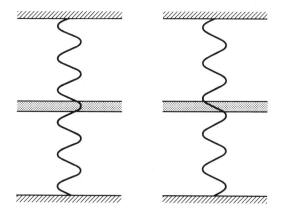

Figure 4.17 Variation of electric field amplitude between plates for adjacent even and odd modes.

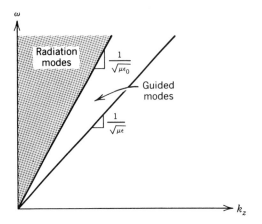

Figure 4.18 Location of radiation and guided modes with respect to the dispersion diagram.

The field solutions for the infinite plate separation are referred to as radiation modes. These modes occupy the shaded portion of the dispersion diagram shown in Figure 4.18 which also indicates, for comparison, the boundaries for propagating guided waves. The complete spatial variation for the radiation-mode electric field is obtained from Eq. 4.40 with Eqs. 4.41 and 4.42 used to eliminate ψ_1 and to obtain the amplitude coefficient A_1 in terms of A_2. The results for $x > 0$ are given by the following expressions:

$d/2 < x < \infty$

$$E_y(x,z) = A_2 \left\{ -\frac{k_{2x}}{k_{1x}} \sin(k_{2x}\, d/2 - \psi) \sin[k_{1x}(x - d/2)] \right.$$
$$\left. + \cos(k_{2x}d/2 - \psi) \cos[k_{1x}(x - d/2)] \right\} e^{-jk_z z}$$

$0 < x \le d/2$

$$E_y(x, z) = A_2 \cos(k_{2x}x - \psi)e^{-jk_z z}$$

Solutions for $x < 0$ are obtained by symmetry.

4.4.2 Extension to the Asymmetric Slab Waveguide (Advanced Topic)

The technique used to obtain the radiation modes for the symmetric slab waveguide can also be used for the more usual geometry in which the waveguide is asymmetric. As with the guided-mode solutions, regions 1, 2, and 3 typically represent an air or cover region, guiding film, and substrate, respectively. With reference to Figure 4.19, it is observed that for this geometry, however, two distinct types of radiation modes will exist in the limit of infinite plate separation. For propagation angles in region 2 that

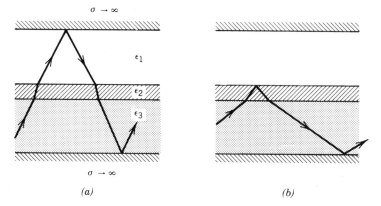

Figure 4.19 Ray path for radiation modes on the asymmetric slab waveguide. (*a*) "Cover" or "air" modes; (*b*) "substrate" modes.

do not exceed the critical angle θ_{cu} for the upper interface, modal solutions will correspond to appropriately oriented propagating waves that travel in a zigzag path between upper and lower conductors. These solutions are analogous to those discussed in the analysis of the symmetric waveguide radiation modes. For the asymmetric waveguide, these solutions are referred to as cover or air radiation modes because an appreciable portion of the total field distribution exists over region 1. Note, however, that for propagation angles within region 2 that exceed the critical angle θ_{cu} at the upper interface, but not the critical angle θ_{cl} at the lower interface the fields in region 1 become evanescent, decaying exponentially as one moves above the upper dielectric interface. These solutions are known as substrate radiation modes because the majority of total field distribution is located within the substrate region. When both θ_{cu} and θ_{cl} are exceeded, the rays are totally internally reflected at both interfaces yielding the guided-wave situation discussed previously. The relationship between k_{2x} and k_z for cover, substrate, and guided modes is indicated in Figure 4.20.

Let us solve for both types of radiation modes for TE wave polarization. Following the same philosophy used in obtaining guided-wave solutions, we initially assume regions 1 through 3 to have different permeabilities as well as permittivities thereby allowing TM solutions to be obtained by duality.

Cover Radiation Modes

We again place our dielectric waveguide between perfectly conducting plates located a distance l_1 above the upper dielectric interface and l_3 below the lower dielectric interface. In analogy with the symmetric guide radiation modes, the cover radiation modes are assumed to be of the form

$$E_y(x, z) = \begin{cases} A_1 \sin[k_{1x}(x - d/2) - \psi_1] \\ A_2 \cos(k_{2x}x - \psi) \\ A_3 \sin[k_{3x}(x + d/2) + \psi_3] \end{cases} e^{-jk_z z} \quad \begin{array}{l} d/2 < x < (l_1 + d)/2 \\ |x| \leq d/2 \\ -(l_3 + d)/2 < x < -d/2 \end{array}$$

$$(4.46)$$

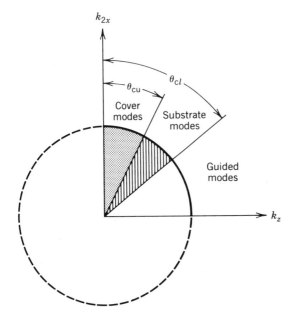

Figure 4.20 Relative propagation angles for radiation and guided modes on the asymmetric slab waveguide. (After Tien (Ref. 11). Reproduced by permission of the author and the Optical Society of America.)

where

$$\psi_1 = k_{1x}l_1 \qquad \psi_3 = k_{3x}l_3$$

Computing tangential **H** from **E** and matching boundary conditions $x = \pm d/2$ leads straight forwardly to the following relations:

$x = +d/2$

$$E_{\text{tan}}: \quad -A_1 \sin \psi_1 = A_2 \cos(k_{2x}d/2 - \psi) \tag{4.47}$$

$$H_{\text{tan}}: \quad A_1 \cos \psi_1 = -\frac{\mu_1}{\mu_2}\frac{k_{2x}}{k_{1x}} A_2 \sin(k_{2x}d/2 - \psi) \tag{4.48}$$

$x = -d/2$

$$E_{\text{tan}}: \quad A_3 \sin \psi_3 = A_2 \cos(k_{2x}d/2 + \psi) \tag{4.49}$$

$$H_{\text{tan}}: \quad A_3 \cos \psi_3 = A_2 \frac{\mu_3}{\mu_2}\frac{k_{2x}}{k_{3x}} \sin(k_{2x}d/2 + \psi) \tag{4.50}$$

Division of Eq. 4.47 by 4.48 and Eq. 4.49 by 4.50 shows that the modal solutions must simultaneously satisfy the two relations:

$$\tan \psi_1 = \frac{\mu_2 k_{1x}}{\mu_1 k_{2x}} \cot(k_{2x}d/2 - \psi) \tag{4.51}$$

$$\tan \psi_3 = \frac{\mu_2 k_{3x}}{\mu_3 k_{2x}} \cot(k_{2x}d/2 + \psi) \tag{4.52}$$

To solve these two expressions, we can represent all the transverse wavenumbers in terms of any single one, say k_{2x}. Then Eqs. 4.51 and 4.52 represent two equations in terms of the two unknowns k_{2x} and ψ. For any choice of plate spacings l_1 and l_3, we obtain a discrete set of allowed values for k_{1x}, k_{2x}, and k_{3x} as well as for ψ. Each solution set $(k_{1x}, k_{2x}, k_{3x}, \psi)$ corresponds to a mode. As with radiation modes on the symmetric waveguide, these solutions simply represent values necessary to fit an integral number of half cycles smoothly between the conductive plates.

The solution to Eqs. 4.51 and 4.52 can be shown to divide into two distinct sets analogous to the even and odd modes of the symmetric waveguide. Specifically, let us demonstrate for any arbitrary solution that we obtain having phase ψ, a second solution always exists having phase $\psi + \pi/2$. To prove this assertion, we first take the inverse cotangent of both sides of Eqs. 4.51 and 4.52, noting that $\cot(x) = \cot(x + \pi)$. Taking the difference of the resulting two expressions and dividing by two then yields

$$\psi = \tfrac{1}{2}\cot^{-1}\left[\frac{\mu_3 k_{2x}}{\mu_2 k_{3x}} \tan(k_{3x}l_3)\right] - \tfrac{1}{2}\cot^{-1}\left[\frac{\mu_1 k_{2x}}{\mu_2 k_{1x}} \tan(k_{1x}l_1)\right] \pm p\,\frac{\pi}{2} \tag{4.53}$$

where

$$p = 0, 1$$

Because the two solutions differ in phase by $\pi/2$, Eq. 4.46 indicates that within region 2 the position of the maxima in electric field for one solution must correspond to the minima for the other. The position of the maxima or minima relative to the center line of the waveguide, $x = 0$, is determined by the value of ψ. Note that from the expression given for ψ in Eq. 4.53, as the symmetric waveguide limit is approached ψ is equal to 0 for $p = 0$ and $\pi/2$ for $p = 1$. In this limit the $p = 0$ solutions have a maximum in electric field amplitude located at the center of the waveguide while the $p = 1$ solutions have a minimum. Since in this limit the $p = 0$ solutions are even and the $p = 1$ solutions are odd, we define the general set of solutions for the asymmetric guide as quasi-even for $p = 0$ and quasi-odd for $p = 1$.

Let us now go directly to the wide plate separation limit. Then by the same arguments made for the symmetric guide the transverse wavenumbers k_{1x}–k_{3x} and the phase ψ make a transition to a continuum of allowed values. Because of the continuum nature of these variables, there are therefore an infinite number of radiation modes. Each mode is specified by selecting a value for any one of the transverse wavenumbers, say

k_{2x}, and a value for the phase ψ. However, as discussed previously, for the cover radiation modes k_{2x} must be chosen in such a fashion that the critical angle is not exceeded at the upper boundary. This implies that

$$\omega\sqrt{\mu_0(\epsilon_2 - \epsilon_1)} < k_{2x} < \infty$$

We observe one significant difference between the solutions for the symmetric and asymmetric cover radiation modes. For the symmetric structure, the phase term ψ was required to be either 0 or $\pi/2$, corresponding to even and odd modes, respectively. However, for the asymmetric guide, the phase can be chosen arbitrarily, being given by some ψ for quasi-even modes and $\psi + \pi/2$ for quasi-odd modes. This result can be understood by making reference to Eq. 4.53. It is observed that the value of ψ depends on both l_1 and l_3. In the limit that the plates are infinitely separated the sum of l_1 and l_3 must approach infinity; however, the difference is not specified. This difference determines how the waveguide is centered with respect to the upper and lower conductive plates and therefore offers an additional degree of freedom over the symmetric waveguide geometry.

Let us examine whether there is any "logical" choice for how the waveguide should be positioned between the plates, thereby uniquely determining ψ. When regions 1 and 3 are identical, symmetry dictates that the physical center of the waveguide should be positioned halfway between the two plates, so that $l_1 = l_3 = l/2$. However, for electromagnetic waves the appropriate unit of measure is not physical length but rather distance measured in wavelengths, or equivalently in cycles. Thus, more importantly, the above choice of plate location for the symmetric guide results in an electric field distribution along x which is centered or electrically "balanced" in that a field maximum or minimum is located exactly halfway between the two plates. Let us define this dividing line as the center of electrical symmetry which in terms of electrical measure, is located an equal number of cycles above and below the conducting plates for all modes. Note that, as is shown by the dashed line in Figure 4.21a, for the symmetric waveguide the physical center of the waveguide corresponds with the electrical center.

Consider now the asymmetric waveguide. For any choice of l_1 and l_3 we calculate from Eq. 4.53 the corresponding phases ψ and $\psi + \pi/2$ for each quasi-even and quasi-odd mode, respectively. The value of ψ determines the shift Δ in the position of the maximum for each quasi-even mode or minimum of each corresponding quasi-odd mode from the line $x = 0$ as shown in Figure 4.21b. The dashed line therefore indicates the electrical center for the asymmetric guide. In analogy with the symmetric structure let us choose l_1 and l_3 in such a fashion that this line is located electrically halfway between upper and lower plates. That is, the total number of cycles from this line to the top plate must equal total number of cycles to the bottom plate. Using this criterion we show in Appendix 1 that ψ is uniquely defined and is given by the relation

$$\tan 2\psi = \left[\frac{1 - \beta}{1 + \beta}\right] \tan(k_{2x}d) \tag{4.54}$$

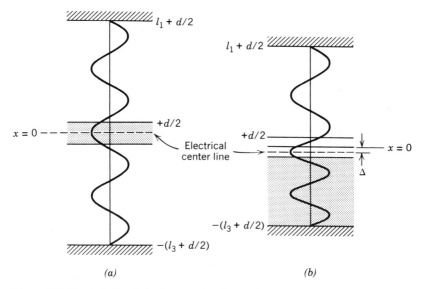

Figure 4.21 Electric field variation between plates for a radiation mode. (a) Symmetric guide; (b) asymmetric guide. The dashed line represents the center of "electrical" symmetry in both cases.

where

$$\beta = \left(\frac{\mu_3}{\mu_1}\right)\left(\frac{k_{1x}}{k_{3x}}\right)\left[1 - \left(\frac{\mu_2}{\mu_3}\right)^2\left(\frac{k_{3x}}{k_{2x}}\right)^2\right]\Bigg/\left[1 - \left(\frac{\mu_2}{\mu_1}\right)^2\left(\frac{k_{1x}}{k_{2x}}\right)^2\right]$$

Note that, unlike the symmetric guide, ψ is not the same for all modes. For the symmetric guide the number of cycles measured along x in regions 1 and 3 is always the same, independent of the value of the transverse wavenumber in these regions. Thus the center of electrical symmetry is fixed for all modes. For the asymmetric guide, however, the number of cycles in regions 1 and 3 in general is not the same and the difference is dependent on the mode. Thus the electrical symmetry center as defined by ψ must change from mode to mode as given in Eq. 4.54.

It should be noted that our convention for choosing ψ has one additional important advantage. It makes our set of quasi-even modal solutions mathematically independent from the quasi-odd set. The consequences of such orthogonality are more fully discussed in Chapter 8.

For infinite plate separation we find, therefore, that for any given value of wavenumber there are two cover radiation modes, one which is "quasi-even" and one which is "quasi-odd." Use of the above expression for ψ along with relations 4.46, 4.51, and 4.52 permits the field-amplitude coefficients A_1 and A_3 to be expressed in terms of A_2 as well as the elimination of the terms containing the plate-spacing-dependent variables ψ_1 and ψ_3. We obtain

Cover-Mode Field Solutions

$\infty > x > d/2$

$$E_y(x, z) = A_2 \left\{ -\frac{\mu_1}{\mu_2}\frac{k_{2x}}{k_{1x}} \sin(k_{2x}d/2 - \psi) \sin[k_{1x}(x - d/2)] \right.$$
$$\left. + \cos(k_{2x}d/2 - \psi) \cos[k_{1x}(x - d/2)] \right\} e^{-jk_z z} \tag{4.55}$$

$|x| \le d/2$

$$E_y(x, z) = A_2 \cos(k_{2x}x - \psi)e^{-jk_z z} \tag{4.56}$$

$-\infty < x < -d/2$

$$E_y(x, z) = A_2 \left\{ \frac{\mu_3}{\mu_2}\frac{k_{2x}}{k_{3x}} \sin(k_{2x}d/2 + \psi) \sin[k_{3x}(x + d/2)] \right.$$
$$\left. + \cos(k_{2x}d/2 + \psi) \cos[k_{3x}(x + d/2)] \right\} e^{-jk_z z} \tag{4.57}$$

with ψ for "quasi-even" modes being defined by expression 4.54 and ψ for "quasi-odd" modes equal to $\psi + \pi/2$. Note that for TE-mode solutions the electric field is given by the above relations with $\mu_1 = \mu_2 = \mu_3$ while TM-mode solutions are obtained from duality by replacing E_y with H_y and $\mu_1 - \mu_3$ with $\epsilon_1 - \epsilon_3$.

Substrate Radiation Modes

Let us solve for the substrate radiation modes directly in the limit that l approaches infinity. The form for the electric field in regions 2 and 3 must be identical to that given for the air radiation modes in expression 4.46, but in region 1 we choose an exponentially decaying field of the form

$$E_y(x, z) = A_1 e^{-\alpha_{1x}x}e^{-jk_z z} \qquad d/2 < x < \infty$$

Computing tangential **H** from **E** and matching boundary conditions at $x = \pm d/2$ yields expressions for the phase shifts ψ and ψ_3:

$$\tan(k_{2x}d/2 - \psi) = \frac{\mu_2}{\mu_1}\frac{\alpha_{1x}}{k_{2x}} \tag{4.58}$$

$$\tan \psi_3 = \frac{\mu_2}{\mu_3}\frac{k_{3x}}{k_{2x}} \cot (k_{2x}d/2 + \psi) \tag{4.59}$$

The field amplitude in region 1 is given by

$d/2 < x < \infty$

$$E_y(x, z) = A_2 \cos(k_{2x}d/2 - \psi)e^{-\alpha_{1x}(x-d/2)}e^{-jk_z z} \tag{4.60}$$

with the field description for regions 2 and 3 identical to those given for the cover modes in expressions 4.56 and 4.57. Again duality may be applied to yield the TM substrate radiation modes.

PROBLEMS

4.1 Assume that the magnetic field for the TM even modes is as given in Eq. 4.5, but with E_y replaced by H_y. Derive the guidance condition for the even TM modes and show how you would obtain a graphical solution.

4.2 Assume that the magnetic field for the TM odd modes is given by Eq. 4.7, but with E_y replaced by H_y. Derive the guidance condition for the odd TM modes and show how you would obtain a graphical solution.

4.3 Determine a general expression for the number of even TE modes that may propagate on a symmetric waveguide having core thickness d and permittivity ϵ. Assume that the surrounding medium is free space and that the frequency of operation is ω.

4.4 Derive Eq. 4.34 from Eq. 4.28.

4.5 A slab dielectric is fabricated by sputtering a thin glass film ($n_2 = 1.620$) on top of a glass substrate ($n_3 = 1.346$).
   (a) Using the b–v diagram of Figure 4.11 determine the maximum film thickness for the guide to propagate only one mode above cutoff. Assume the light source is a He–Ne laser ($\lambda_0 = 0.63$ μm in free space).
   (b) What is the bounce angle for the fundamental mode when the film is this thickness?

4.6 Show that for highly asymmetric waveguides ($a \to \infty$) the guidance condition reduces to

$$- \cot(k_{2x}d) = \alpha_{3x}/k_{2x}$$

Hint. Use Eq. 4.34.

4.7 Show that for the highly asymmetric waveguide the refractive-index difference between the film and substrate, $\Delta n = n_2 - n_3$, required for the pth mode to propagate is given by

$$\Delta n \cong \frac{(2p + 1)^2}{32n_3} \left(\frac{\lambda_0}{d}\right)^2$$

where λ_0 is the free-space wavelength.

4.8 A symmetric slab waveguide has a core dimension $d = 3$ μm with a refractive index $n_2 = 1.5$. Assume the surrounding medium is free space and that the waveguide is excited with a He–Ne laser having a free space $\lambda_0 = 0.63$ μm.
(a) How many TE modes, even plus odd, can the waveguide support?
(b) What TE mode has the lowest cutoff frequency and what is f_{cp} for this mode?
(c) What is the cutoff frequency for the next highest TE mode?
(d) Suppose that instead of the region surrounding being free space it is a dielectric having $n_1 = 1.48$. How many modes can now propagate?

4.9 Design a symmetric slab waveguide by choosing the core thickness d such that the bounce angle for the fundamental mode is 87°. Your given material parameters are

$$n_2 = 1.51$$
$$n_1 = 1.46$$
$$\lambda_0 = 0.63 \text{ μm}$$

4.10 A symmetric slab waveguide has the following parameters: $d/\lambda_0 = 0.9$, $\lambda_0 = 0.63$ μm, $n_2 = 1.5$, $n_1 = 1.48$. What is the effective waveguide thickness for the fundamental mode?

4.11 Compute the effective dielectric constant for the first three TE modes on the asymmetric waveguide having the following specifications: $n_1 = 1.21$, $n_2 = 1.45$, $n_3 = 1.43$, $d/\lambda_0 = 5.3$.

4.12 (a) Show that d_{eff} for TE modes on the asymmetric slab guide can be written as

$$d_{eff} = d\{1 + (1/v)[1/(b)^{1/2} + 1/(b + a)^{1/2}]\}$$

(b) Make a plot of d_{eff} versus v for $a = 0$ and $a = \infty$.

4.13 A symmetric waveguide has $v = 10$ and $a = 0$. Find b for the modes $p = 0$, 1, and 2 from Figure 4.11a. Verify the correctness of these solutions by substitution into Eq. 4.34.

4.14 Show that the generalized guidance condition for TE modes, Eq. 4.34, yields two sets of solutions in the special case for which the guide is symmetric ($a = 0$). Compare these solutions with the guidance conditions, Eqs. 4.6 and 4.8, for the TE mode symmetric waveguide.

REFERENCES

1. Barnoski, M. K. *Introduction to Integrated Optics*. New York: Plenum, 1973.
2. Hunsperger, R. G. *Integrated Optics: Theory and Technology*. New York: Springer-Verlag, 1982.
3. Kapany, N. S., and Burke, J. J. *Optical Waveguides*. New York: Academic, 1972.

4. Kogelnik, H. "An introduction to integrated optics." *IEEE Transactions on Microwave Theory and Techniques* MTT-23 (1975): 2–20.

5. Kogelnik, H., and Ramaswamy, V. "Scaling rules for thin-film optical waveguides." *Applied Optics* 13 (1974): 1857–1862.

6. Marcuse, D. *Theory of Dielectric Waveguides.* New York: Academic, 1974.

7. Marcuse, D. *Light Transmission Optics.* New York: Van Nostrand-Reinhold, 1972.

8. Shevchenco, V. V. *Continuous Transitions in Open Waveguides.* Boulder, Colorado: Golem Press, 1971.

9. Tamir, T., ed. *Integrated Optics.* New York: Springer-Verlag, 1975.

10. Taylor, H. F., and Yariv, A. "Guided wave optics." *Proceedings of the IEEE* (1974): 1044–1060.

11. Tien, P. K. "Light waves in thin films and integrated optics." *Applied Optics* 10 (1971): 2395–2412.

CHAPTER 5

Practical Waveguiding Geometries

5.1 INTRODUCTION

The asymmetric slab dielectric waveguide analyzed in the last chapter is an accurate representation of the type of guiding structure that is formed, for example, when a thin guiding film of high permittivity is deposited upon a slightly lower permittivity substrate. For this geometry, the permittivity is constant, except for an abrupt discontinuity at the air–film and film–substrate interfaces. Such a variation in ϵ with transverse coordinate x is referred to as piecewise continuous and is illustrated in Figure 5.1a.

A number of the techniques presently used in fabricating practical optical integrated circuit (OIC) waveguides, however, result in a significantly different type of guiding structure. When ions, atoms, or molecules are driven or diffused into or out of the surface of a substrate, the dielectric properties of the surface region are modified, creating a top "layer" of high permittivity. This modified layer serves a purpose analogous to the guiding film in the slab waveguide. Unlike the slab waveguide, however, the permittivity of the guiding "layer" formed by such a process exhibits a smooth, continuous variation with increasing depth into the substrate as shown in Figure 5.1b. Such a structure is referred to as a graded-index waveguide and is analyzed in Section 5.2. Fabrication methods for producing graded-index guides are discussed in detail in Chapter 7.

Another important feature of many OIC waveguides is the capability to confine optical energy in both transverse directions, x and y. This allows the size of the OIC components to be reduced so that they may be placed in proximity on the same wafer. When confinement occurs in both transverse directions, the waveguide is often referred to as a three-dimensional structure because, in contrast to the two-dimensional structures studied so far, field variations in three rather than two directions must be considered. A number of three-dimensional waveguiding structures exist including the raised or embedded strip, rib or ridge, and strip-loaded waveguides. These basic waveguides are shown in Figure 5.2.

For the strip or embedded-strip waveguide, transverse confinement in both the x and y directions can be easily explained because there exists a high permittivity or core region ϵ_2 which is completely surrounded by regions of lower dielectric constant, ϵ_1 and ϵ_3. Thus, in terms of the bouncing beam concept, guidance can be considered to be the result of rays which propagate in a zigzag fashion within the high-permittivity

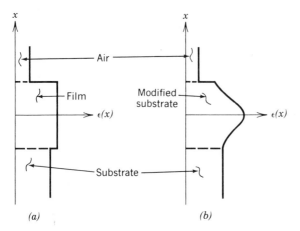

Figure 5.1 Permittivity profile. (*a*) Step-index guide; (*b*) graded-index guide.

region, suffering total internal reflection at the ϵ_2–ϵ_1 and ϵ_2–ϵ_3 interfaces. The path followed by these rays is shown schematically in Figure 5.3.

Unfortunately, the above interpretation of guidance in the strip guide is not easily extended to analyze the rib and strip-loaded guides because no central high-permittivity region capable of confinement in both x and y directions can be identified. An alternative

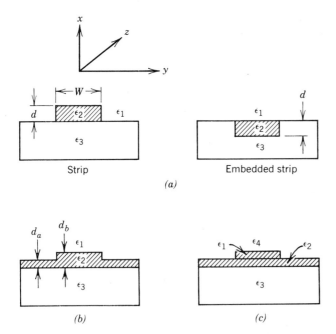

Figure 5.2 Various types of waveguide geometries. (*a*) Strip and embedded strip, $\epsilon_2 > \epsilon_3 > \epsilon_1$; (*b*) rib or ridge, $\epsilon_2 > \epsilon_3 > \epsilon_1$; (*c*) strip loaded, $\epsilon_2 > (\epsilon_1, \epsilon_3) > \epsilon_4$.

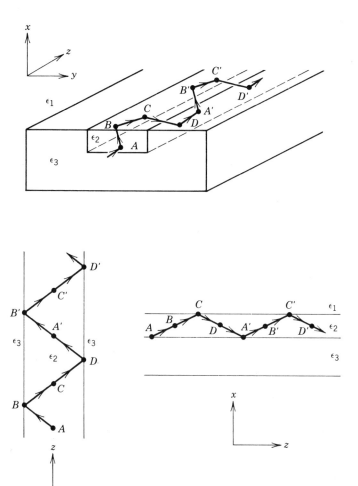

Figure 5.3 Ray path for embedded strip waveguide.

viewpoint, however, known as the effective index method, can be used in a unified fashion to explain confinement along y for all of the waveguiding structures shown in Figure 5.2. The essence of this technique may be understood by observing that for all of these structures, the permittivity, when averaged along the $\hat{x}$ direction, is greater for the central region $|y| < W/2$ than for the surrounding regions $|y| > W/2$. Thus, in terms of confinement in the $\hat{y}$ direction, these guides may be viewed as ''equivalent'' to a fictitious symmetric slab guide consisting of a central region of high effective or average permittivity for $|y| < W/2$ sandwiched between two identical lower effective permittivity regions occupying the regions $|y| > W/2$. The method for computing the effective permittivities and obtaining the guidance condition is described in Section 5.3, and makes use of an extension of the analysis developed for the two-dimensional step-index guide in Chapter 4.

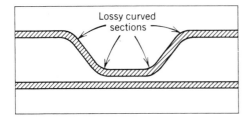

Figure 5.4 Top view of waveguide showing curved portion.

A third way in which the structure of practical OICs deviates from that of the ideal waveguides studied so far is in the inclusion of curved guiding sections as shown in Figure 5.4. These curved sections are required to permit the interconnection of the various optical or electro-optical components located on the chip. As a result of these bends, a portion of the confined radiation invariably leaks out of the guide, resulting in attenuation of the optical signal. The cause of this radiation loss and its analysis is presented in Section 5.4.

5.2 GRADED-INDEX TWO-DIMENSIONAL WAVEGUIDES: THE WKB METHOD

As discussed previously, waveguide fabrication techniques often result in waveguiding structures for which the permittivity varies smoothly in a plane transverse to the direction of propagation. This type of guiding structure is in contrast with the slab waveguide geometry, which consists of three piecewise continuous dielectric regions. For the slab geometry, field solutions satisfying Maxwell's equations may be found straightforwardly for each uniform dielectric region and subsequently matched across the interfaces to obtain the allowed modes. This technique, however, is not applicable to geometries exhibiting a continuous variation in permittivity with transverse spatial coordinates. Instead, a powerful mathematical technique known as the WKB method is presented that allows the propagation constant and transverse field variation to be obtained approximately for each guided mode.

5.2.1 Formulation

Let us consider a dielectric waveguide having the variation in permittivity along x shown in Figure 5.1b. We assume the guiding region is imbedded sufficiently deeply within the substrate so that the influence of the dielectric discontinuity at the substrate–air interface can be ignored. With reference to Figure 5.1b, we observe that from a qualitative point of view, the nonuniform waveguide is similar in structure to the dielectric slab guide, having a guiding "region" of high permittivity surrounded by two lower permittivity regions. The boundaries of the guiding region are nebulous at this point, but if these can be found then it is expected that the fields should exhibit an oscillatory behavior within this region and decay away exponentially to either side.

To put the analysis on more quantitative footing, an analogy is made with the guidance condition developed for the slab waveguide using ray theory. Therefore, let us consider the influence of the inhomogeneous permittivity on a ray propagating within the waveguide. It is assumed that locally, at some point (x, z) along the guide, the ray "sees" a permittivity $\epsilon(x)$ and acts locally as a plane wave satisfying the dispersion relation,

$$k_x^2(x) + k_z^2 = \omega^2 \mu \epsilon(x) \equiv k^2(x) \qquad (5.1)$$

Each ray is assumed to be one out of an infinite number of identical beams composing a guided mode. Thus, each ray must have the same longitudinal wavenumber $(k_z)_p$ as the guided mode of which it is a portion. As with the slab waveguide, $(k_z)_p$ is assumed to take on a discrete set of allowed values, one for each guided mode. These values are yet to be determined, however. Equation 5.1 can be rearranged to solve for the value of the transverse component of the wave vector k_x:

$$k_x(x) = k_0 \sqrt{n^2(x) - n_{\text{eff}}^2} \qquad (5.2)$$

where

$$n^2(x) = \epsilon(x)/\epsilon_0$$

and

$$n_{\text{eff}}^2 = k_z^2/k_0^2$$

For any given mode, the propagation angle of the ray $\theta(x)$ measured with respect to the $\hat{z}$ axis is obtained from the relation

$$k(x) \cos \theta(x) = k_z$$

or

$$\theta(x) = \cos^{-1}[n_{\text{eff}}/n(x)] \qquad (5.3)$$

Thus, the ray propagates at an angle that is dependent upon its x coordinate. We observe that the ray is most inclined with respect to the z axis at those points where $n(x)$ is a maximum and that it propagates parallel to z where $n(x) = n_{\text{eff}}$. The path of the ray is shown schematically in Figure 5.5 where we have defined the points x_A and x_B to correspond to the positions along x where the ray is parallel to the z axis. The quantities x_A and x_B are known mathematically as the turning points. From Eqs. 5.2 and 5.3, the turning points are observed to occur where

$$k_x(x_A) = k_x(x_B) = 0$$

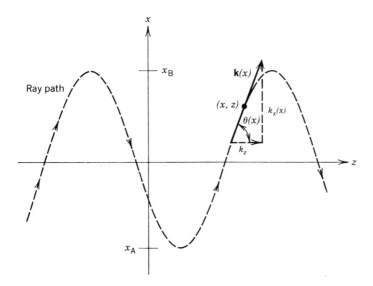

Figure 5.5 Generalized ray path for graded-index waveguide. Permittivity is assumed to vary along x.

or equivalently where

$$n(x_A) = n(x_B) = n_{eff} \tag{5.4}$$

Equation 5.4 is represented diagramatically in Figure 5.6. For $x < x_A$ or $x > x_B$ we see from Eq. 5.2 that k_x makes a transition from purely real to purely imaginary, indicating that the fields decay exponentially beyond this point just as they do outside the core region of the slab guide. This gives additional physical interpretation to the turning points as the equivalent "boundaries" for the core of the dielectric guide.

To obtain the value of n_{eff} explicitly for each mode, analogy is made with the slab waveguide. Recall that for this structure, the guidance condition requires that the net phase shift in the $\hat{x}$ direction incurred by a ray over a single round-trip bounce between dielectric "interfaces" must be an integral multiple of 2π. For a waveguide having a core thickness d this relation was shown to be

$$-2k_{2x}d + \phi_1 + \phi_3 = 2p\pi \tag{5.5}$$

Here, $-2k_{2x}d$ represents the phase shift along x incurred due to propagation, and ϕ_1 and ϕ_3 represent the additional phase shifts resulting from total internal reflection at the two dielectric interfaces.

For the inhomogeneous wave, the differential phase shift along x incurred due to the ray traveling from x to $x + dx$ is

$$d\phi = -k_x(x)\, dx$$

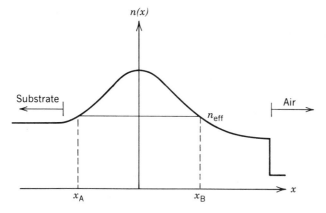

Figure 5.6 Variation in refractive index with x, location of turning points, and value of effective index n_{eff}.

Thus, the round-trip phase shift from x_A to x_B and back is

$$\phi = -2 \int_{x_A}^{x_B} k_x(x) \, dx$$

We need to show that at the turning points x_A and x_B phase shifts due to "reflection" are incurred in a fashion analogous to that in the slab guide. Recall that the reflection coefficient for TE and TM waves incident from region 2 onto region 1 are given, respectively, by

$$R^{TE} = \frac{1 - k_{1x}/k_{2x}}{1 + k_{1x}/k_{2x}}$$

and

$$R^{TM} = \frac{1 - (\epsilon_2 k_{1x})/(\epsilon_1 k_{2x})}{1 + (\epsilon_2 k_{1x})/(\epsilon_1 k_{2x})}$$

Let us write an equivalent expression for the inhomogeneous guide, treating the turning points x_A and x_B as effective boundaries. We obtain at x_A for both TE and TM polarizations,

$$R_A = \frac{1 - k_x(x_A^-)/k_x(x_A^+)}{1 + k_x(x_A^-)/k_x(x_A^+)}$$

where

$$x_A^{\pm} \equiv x_A \pm \lim_{\delta x \to 0} (\delta x)$$

and δx is a small positive increment in distance from x_A. The reflection coefficient is the same for TE and TM polarizations because $\epsilon(x)$ is continuous as we go across the "boundary" at $x = x_A$. As shown in Figure 5.6 near $x = x_A$, the refractive index $n(x)$ can be expanded to first order as

$$n(x) = n_{eff} + (x - x_A) \Delta n/\Delta x \qquad (5.6)$$

Here $\Delta n/\Delta x$ represents the slope of $n(x)$ near x_A. As will be seen, its explicit value is unimportant. Thus, the transverse wavenumber $k_x(x)$ when evaluated near x_A is given from Eqs. 5.2 and 5.6 by

$$k_x(x) \cong k_0 \sqrt{2 (x - x_A)(\Delta n/\Delta x)n_{eff}}$$

Substituting $x_A^{\pm}$ into the above relation for $k_x(x)$ therefore yields

$$k_x(x_A^+) = \delta k_x$$

and

$$k_x(x_A^-) = j \, \delta k_x$$

where

$$\delta k_x \equiv \lim_{\delta x \to 0} k_0 \sqrt{2\delta x(\Delta n/\Delta x)n_{eff}}$$

The reflection coefficient at $x = x_A$ is therefore equal to

$$R_A = \frac{1 - j}{1 + j} = 1 \, e^{-j\pi/2}$$

which has unity magnitude and introduces a phase shift of $\phi_A = -\pi/2$. A similar analysis at $x = x_B$ shows that $|R_B| = |R_A|$ and $\phi_B = \phi_A$. Thus the guidance condition becomes

$$-2 \int_{x_A}^{x_B} k_x(x) \, dx + \phi_A + \phi_B = 2p\pi$$

or

$$\qquad (5.7)$$

$$k_0 \int_{x_A}^{x_B} \sqrt{n^2(x) - n_{eff}^2} \, dx = (p + \tfrac{1}{2})\pi \qquad p = 0, 1, 2, \ldots$$

with x_A and x_B defined implicitly through the relations

$$n(x_A) = n(x_B) = n_{eff}$$

Equation 5.7 is analogous to the guidance condition for the dielectric slab. It states that the value of n_{eff} cannot be chosen arbitrarily but must be of such a value that the integrated phase given by the left side of Eq. 5.7 must equal an odd multiple of $\pi/2$. Each choice of the integer p defines a different mode and a corresponding value of $(n_{eff})_p$. As with the slab guide, p represents the number of half "cycles" of field variation along the x direction between the turning point "boundaries."

In practice, Eq. 5.7 must in general be solved numerically by trial and error. First, a value for the integer p is chosen, thereby selecting the mode of interest. Next, a guess for the value of n_{eff} is made. This also defines the values for x_A and x_B through Eq. 5.4. The left side of Eq. 5.7 can then be evaluated numerically and compared with the right side. The process is repeated until convergence is obtained or until it is found that no solution is possible. As discussed below, the latter alternative implies that the mode is cut off.

A number of observations can be made concerning mode behavior by examining the properties of Eq. 5.7. Consider first the relation between the mode index p and the corresponding effective refractive index for this mode, $(n_{eff})_p$. As we go to increasingly higher order modes, that is, larger values of the integer p, the right side of Eq. 5.7 increases. The left side of the equation must, of course, increase by a corresponding amount. This is accomplished by making the associated value of $(n_{eff})_p$ smaller, thereby increasing both the size of the integrand and the range of integration. Thus, higher order modes have increasingly smaller values of $(n_{eff})_p$. This relation is shown in Figure 5.7. Note that the separation between turning points increases with increasing mode index so that more half cycles of field variation fit between. From this figure it is also clear that only a limited number of modes, $p_{max} + 1$, can propagate at any given frequency. For modes higher than $p = p_{max}$, $(n_{eff})_p$ is sufficiently small that either one or both of the turning points do not exist. These higher order modes cannot propagate because they are unable to be totally internally reflected.

To obtain the cutoff frequency, f_{cp}, for the pth mode reference is again made to Eq. 5.7. If p is fixed and ω, or equivalently k_0, is reduced, $(n_{eff})_p$ must also be reduced to maintain the equality in this expression. Our solution for $(n_{eff})_p$ is lost when

$$(n_{eff})_p = n_{min}$$

where n_{min} is the value of $n(x)$ at which a turning point no longer exists, as shown in Figure 5.7. Thus, from Eq. 5.7, setting $n_{eff} = n_{min}$ and solving for f_{cp}, we obtain

$$f_{cp} = \frac{c(p + \frac{1}{2})}{2 \int_{x_A}^{x_B} \sqrt{n^2(x) - n_{min}^2} \, dx}$$

The turning points at the cutoff frequency are defined implicitly through the relation

$$n(x_{A,B}) = n_{min}$$

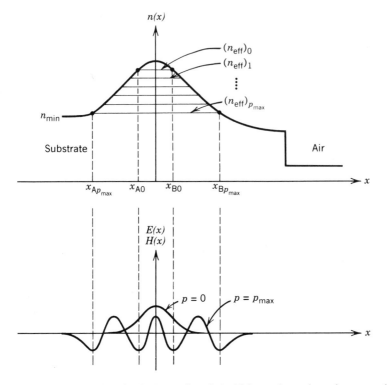

Figure 5.7 Value of n_{eff} for the lowest through the highest order modes and corresponding field variation for the lowest and highest order modes.

Example 5.1

To demonstrate how $(n_{eff})_p$ can be obtained from Eq. 5.7 by numerical iteration, we consider a waveguide having the following variation in refractive index:

$$n^2(x) = n_{max}^2 [1 - (x/x_0)^2] \qquad |x| < x_0/2$$

$$= n_{min} \qquad\qquad |x| > x_0/2$$

Provided that $(n_{eff})_p > n_{min}$ (that is, the mode is not cut off) we must solve the equation

$$\frac{2\pi}{\lambda_0} \int_{x_A}^{x_B} \sqrt{(n_{max}^2 - n_{eff}^2) - (n_{max} x/x_0)^2} \, dx - (p + \tfrac{1}{2}) \pi = 0$$

where

$$x_{A,B} = \mp x_0 \sqrt{1 - (n_{eff}/n_{max})^2}$$

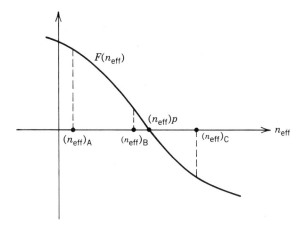

Figure 5.8 Graphical demonstration of a method for obtaining the solution to $F(n_{\text{eff}}) = 0$.

The above relation is of the form

$$F(n_{\text{eff}}) = 0$$

which is shown diagramatically in Figure 5.8 for the region near some solution, $n_{\text{eff}} = (n_{\text{eff}})_p$.

The value of $(n_{\text{eff}})_p$ can be obtained by first choosing two guesses for (n_{eff}), say $(n_{\text{eff}})_A$ and $(n_{\text{eff}})_C$. $F(n_{\text{eff}})$ is then evaluated for these two guesses. Provided $F(n_{\text{eff}})$ is of opposite sign for the two initial guesses, then $(n_{\text{eff}})_p$ must lie within the interval defined by these two points. The interval is now halved to yield a new point, $(n_{\text{eff}})_B = (n_{\text{eff}})_A + \frac{1}{2}[(n_{\text{eff}})_C - (n_{\text{eff}})_A]$, as shown in Figure 5.8. Again, $F(n_{\text{eff}})$ is tested to see whether it changes sign between $(n_{\text{eff}})_A$ and $(n_{\text{eff}})_B$ or between $(n_{\text{eff}})_B$ and $(n_{\text{eff}})_C$. The appropriate new interval is selected and the process repeated until the interval about $(n_{\text{eff}})_p$ is as small as is desired. Because the interval is successively halved, after N iterations the uncertainty in the value of $(n_{\text{eff}})_p$ is reduced by a factor of 2^N over the initial interval. This process is shown in Table 5.1 for $x_0 = \lambda_0$, $n_{\text{max}} = 1.5$ and $p = 0$. The top numbers represent the guesses for n_{eff} and the numbers in parentheses represent $F(n_{\text{eff}})$ for each guess. The exact solution, obtained in the next section, is given by $(n_{\text{eff}})_p = 1.4182$ which is within 0.04% of the guess obtained after only four iterations.

5.2.2   Guidance by a Parabolic Variation in Permittivity

While the guidance condition, Eq. 5.7, in general must be solved numerically, one very important refractive index profile, for which the refractive index varies parabolically with x, can be solved in closed form. This profile can be useful as an approximation for guides that exhibit a more complicated variation for $\epsilon(x)$ but which are parabolic over some reasonable distance. Figure 5.9 shows a parabolic fit made to

Table 5.1 Demonstration of Numerical Convergence to the Value of n_{eff}

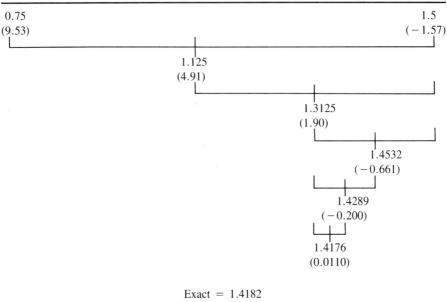

Exact = 1.4182

some arbitrary permittivity profile. We observe that, provided the field solutions found from WKB theory using the parabolic profile do not have appreciable amplitude in the regions where the parabolic fit diverges significantly from the true profile, $\epsilon(x)$, then we should obtain a good approximation to the true solution. To put this constraint into mathematical terms, reference is made to Figure 5.9. The origin for the x axis is chosen at the peak value of the waveguide permittivity. We let the distance L be defined as the maximum positive or negative excursion measured from the origin for which the

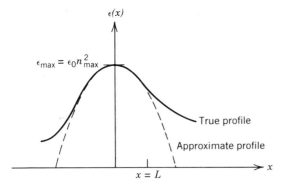

Figure 5.9 Parabolic approximation of the true permittivity profile.

parabolic approximation to $\epsilon(x)$ and the true permittivity profile fit closely. We require that the distance from the origin to both turning points x_A and x_B be much less than L. Since the turning points give locations of the beginning of the evanescent tails for each guided mode they are, therefore, a rough measure of the spatial extent of each mode.

With the above constraints in mind, let us represent $\epsilon(x)$ approximately as

$$\epsilon(x) \cong \epsilon_{max} [1 - (x/x_0)^2] \tag{5.8}$$

The quantity x_0 determines the curvature of the parabola. From Eq. 5.7, the guidance condition becomes

$$k_0 \int_{x_A}^{x_B} \sqrt{(n_{max}^2 - n_{eff}^2) - n_{max}^2 (x/x_0)^2} \, dx = (p + \tfrac{1}{2})\pi \tag{5.9}$$

The turning points x_A and x_B are defined as the values of x for which the integral above goes to zero. Thus

$$x_{A,B} = \mp x_0 \, \delta \tag{5.10}$$

where

$$\delta^2 \equiv 1 - (n_{eff}/n_{max})^2 \tag{5.11}$$

Substituting these integration limits into Eq. 5.9 yields

$$k_0 \, n_{max} \int_{-x_0\delta}^{x_0\delta} \sqrt{\delta^2 - (x/x_0)^2} \, dx = (p + \tfrac{1}{2})\pi \tag{5.12}$$

Let us make the change of variables

$$x = x_0\delta \sin \theta$$

so that

$$dx = x_0\delta \cos \theta \, d\theta$$

Then Eq. 5.12 becomes

$$k_0 \, n_{max} \, \delta^2 \, x_0 \int_{-\pi/2}^{\pi/2} \cos^2 \theta \, d\theta = (p + \tfrac{1}{2})\pi$$

which after integration yields

$$k_0 \, n_{max} \, \delta^2 \, x_0 \, \pi/2 = (p + \tfrac{1}{2})\pi$$

The above result can be solved for n_{eff} using Eq. 5.11 yielding

$$n_{eff}^2 = n_{max}^2 - \frac{(2p + 1)n_{max}}{k_0 x_0} \qquad (5.13)$$

To insure the validity of the above solution, we must require that the magnitude of the turning points is much less than L. From Eqs. 5.10, 5.11, and 5.13, the turning points $x_{A,B}$ are

$$x_{A,B} = \mp \left[\frac{(2p + 1)}{k_0\, n_{max}\, x_0} \right]^{1/2} x_0$$

so that the constraint $|x_{A,B}| \ll L$ can be written as

$$p \ll (\tfrac{1}{2} k_0\, n_{max}\, x_0)\, (L/x_0)^2 - \tfrac{1}{2}$$

5.2.3 Ray Trajectory

Figure 5.5 indicated qualitatively the expected ray trajectory for the guided modes on an inhomogeneous waveguide. Let us examine in a quantitative fashion the path followed by the guided waves in the parabolic approximation for the refractive index. We will consider, for simplicity, so-called paraxial modes, which, by definition, have ray trajectories that are nearly along the propagation axis z. Note that since from Eq. 5.3

$$\theta(x) = \cos^{-1}[n_{eff}/n(x)]$$

then $\theta(x)$ is a maximum where $n(x)$ is largest, that is, where $n(x) = n_{max}$. Thus, a modal solution p will be paraxial *for all* x provided that its value of $(n_{eff})_p$ is approximately equal to n_{max}.

To obtain the ray trajectory explicitly, reference is made to Figure 5.5. Consider the arbitrary point (x, z) on the path of a paraxial ray.

The slope of the path at this point for paraxial rays is described by

$$\frac{dx(z)}{dz} = \frac{k_x(x)}{k_z} = \frac{k_x(x)}{k_0\, n_{eff}} \cong \frac{k_x(x)}{k_0\, n_{max}}$$

From Eqs. 5.2 and 5.8

$$k_x(x) = k_0\, n_{max} \sqrt{\delta^2 - (x/x_0)^2} \qquad (5.14)$$

so that

$$\frac{dx}{dz} \cong \delta \sqrt{1 - (x/x_0\delta)^2} \qquad (5.15)$$

Let us make the substitution of variables

$$\frac{x}{x_0 \delta} = \sin x' \qquad (5.16)$$

so that

$$dx = x_0 \delta \cos x' \, dx'$$

Then Eq. 5.15 simplifies to

$$\frac{dx'}{dz} = \frac{1}{x_0}$$

or

$$x' = \frac{z}{x_0} + x_0'$$

where x_0' is a constant. Therefore, from Eq. 5.16,

$$x = x_0 \delta \sin\left(\frac{z}{x_0} + x_0'\right)$$

Let us choose our z origin such that $x(z = 0) = 0$, thus $x_0' = 0$. Therefore, making use of Eqs. 5.11 and 5.13 to describe the parameter δ in terms of the mode index p yields for the ray trajectory

$$x(z) = x_0 \left[\frac{(2p + 1)}{k_0 \, n_{max} \, x_0}\right]^{1/2} \sin(z/x_0) \qquad (5.17)$$

It is of interest to examine the ray trajectory as a function of mode number p. We note that the peak value of $x(z)$ increases with increasing mode number but that the periodicity is the same for all paraxial modes. This is in contrast with the modes of the slab waveguide in which the peak value of x is fixed by the boundaries of the core for all modes, but where the pitch of the zigzag path is a function of p. This result proves to be important in the analysis of intermodal dispersion for optical fibers discussed in Chapter 10.

5.3 THREE-DIMENSIONAL WAVEGUIDING

While the slab dielectric waveguide provides a means for analyzing most of the important characteristics of guided optical radiation, it is not a very useful geometry for practical applications because of the infinite spatial extent of the wave in one direction

($\hat{y}$). As was discussed in the introduction to this chapter, a number of geometries may be employed to create transversal confinement of the wave both along $\hat{x}$ and $\hat{y}$. In this section, the methods developed for the step index two-dimensional waveguide will be extended to analyze the three-dimensional (3-D) waveguide.

5.3.1   Effective Index Method for the Embedded-Strip Waveguide

Let us consider for analysis the embedded-strip 3-D waveguide shown in Figure 5.2*a*. This guide consists of a rectangular region or core of high permittivity ϵ_2, embedded in a substrate of lower permittivity ϵ_3. The surrounding region above is assumed to have permittivity ϵ_1, which is usually equal to ϵ_0. This structure may be considered to be an idealization of the type of waveguide that can be formed by diffusion, as described in Chapter 7. Although we limit our analysis in this section to this one specific 3-D waveguide, it will be shown in the next section how the methods developed herein can be extended to analyze the other various types of 3-D guides.

Guidance for this structure can be viewed as the result of total internal reflection at the ϵ_2–ϵ_1, and ϵ_2–ϵ_3 interfaces. Because, for practical waveguides, the core permittivity in region 2 is only slightly greater than that of the surrounding region 3, the bouncing wave must propagate nearly parallel to the z axis for total internal reflection to occur. The propagation wave vectors for these bouncing waves must, therefore, be directed nearly along $\hat{z}$. Thus, from Maxwell's equations, the fields **E** and **H** for each mode are perpendicular to **k** and therefore nearly transverse to $\hat{z}$. Assuming that two orthogonal polarizations are possible, we therefore can have either $\mathbf{E} \cong \hat{y}E_y$ and $\mathbf{H} \cong \hat{x}H_x$ or alternatively $\mathbf{E} \cong \hat{x}E_x$ and $\mathbf{H} \cong \hat{y}H_y$.

If the waveguide is viewed in the xz plane, as shown in Figure 5.10*a*, then in the region $|y| < W$ the structure appears identical to that of an asymmetric slab waveguide having a guiding region of permittivity ϵ_2 and width d, as shown. One may be tempted, therefore, to proceed with the analysis in precisely the same fashion as for the slab waveguide, obtaining the propagation constant k_z from the normalized b–v dispersion curves given in Chapter 4. However, matters are not quite this simple because to use these solutions would require that the fields do not vary with y. This, unfortunately, is not the case because the guiding structure also appears as a symmetric slab waveguide when viewed in the yz plane, as shown in Fig. 5.10*b*. Thus, since waveguiding occurs in both the $\hat{x}$ and $\hat{y}$ directions, our fields must vary with both x and y.

Suppose we forget about the field variation with respect to y momentarily, or equivalently assume W is very large. The solutions to this problem would, therefore, correspond to finding the allowed guided modes on the asymmetric slab shown in Figure 5.10*a*. If the assumed polarization of the electric field is along $\hat{y}$, then the solutions correspond to modes which are TE with respect to the xz plane. Alternatively, if **E** is along $\hat{x}$ then **H** is along $\hat{y}$ and the solutions are TM with respect to this plane. For any given frequency, the propagation constants for these modes can be obtained directly from the normalized dispersion curves derived in Section 4.3.4. Specifically, given ω and values for the waveguide dimensions and various permittivities, then we

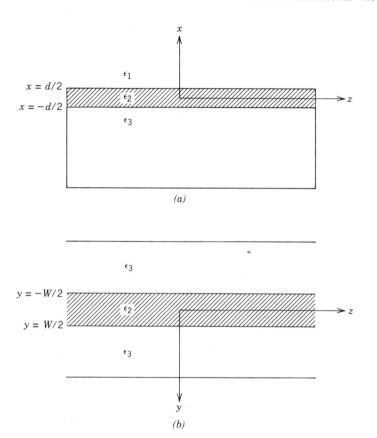

Figure 5.10 Geometry for a three-dimensional waveguide. (*a*) View in the *xz* plane for $|y| < W/2$; (*b*) view in the *yz* plane for $|x| < d/2$.

can compute the normalized frequency parameter

$$v = k_0 d (n_2^2 - n_3^2)^{1/2} \tag{5.18}$$

and either the asymmetry constant a^{TE} or a^{TM}. Here n_2 and n_3 represent the refractive indices in regions 2 and 3, respectively. In the usual case for which $(n_2 - n_3)/n_2$ is small, then, as discussed in Chapter 4, the same normalized dispersion curve can be used for both TE and TM modes. From this curve we obtain a solution for the value of the parameter b for each allowed mode p, where b_p is defined by

$$b_p = [(n_{\text{eff}})_p^2 - n_3^2]/(n_2^2 - n_3^2) \tag{5.19}$$

The index, $p = 0, 1, \ldots$, indicates the number of half cycles of field variation along x in the core region, 2. For each value of b_p, there corresponds an effective refractive index $(n_{\text{eff}})_p$. The propagation constant $(k_z)_p$, associated with each solution, is related to the effective refractive index by the definition

$$(k_z)_p = k_0(n_{\text{eff}})_p \qquad (5.20)$$

Note that this represents the propagation constant in the limit that W is infinite.

The corresponding transverse wavenumbers in regions 1, 2, and 3, $(\alpha_{1x})_p$, $(k_{2x})_p$, and $(\alpha_{3x})_p$ can now be obtained by using the appropriate dispersion relation in each respective region. These are given by

$$(\alpha_{1x})_p = \sqrt{k_z^2 - k_1^2} = k_0\sqrt{(n_{\text{eff}})_p^2 - n_1^2}$$

$$(k_{2x})_p = \sqrt{k_2^2 - k_z^2} = k_0\sqrt{n_2^2 - (n_{\text{eff}})_p^2}$$

and

$$(\alpha_{3x})_p = \sqrt{k_z^2 - k_3^2} = k_0\sqrt{(n_{\text{eff}})_p^2 - n_3^2}$$

Thus, the entire information about confinement in the xz plane is contained in the quantity $(n_{\text{eff}})_p$. Therefore, in terms of the propagation characteristics of our modes, in the region $|y| < W/2$ the entire asymmetric slab could be replaced by an equivalent uniform single material of permittivity $(\epsilon_{\text{eff}})_p$. Note that because of the subscript p, each mode would in general require a different value for this effective index. This equivalent replacement yields the fictitious waveguide shown in Figure 5.11.

We now use our substituted effective permittivity to analyze the guidance in the yz plane. The fictitious waveguide appears as a symmetric structure having core permittivity $(\epsilon_{\text{eff}})_p$ of width W surrounded by permittivity ϵ_3. Since (ϵ_{eff}) is known, then we can calculate a new parameter v' which is analogous to v:

$$v' = k_0W[(n_{\text{eff}})_p^2 - n_3^2]^{1/2} = vb_p^{1/2}W/d \qquad (5.21)$$

Note that since the structure is symmetric in the yz plane, the asymmetry measure in this plane, a', is equal to zero. Again, based on the weak-guiding assumption, we can use the normalized dispersion curve for TE modes to compute a value for the parameter b' from v':

$$(b')_{pq} = [(n'_{\text{eff}})_{pq}^2 - n_3^2]/[(n_{\text{eff}})_p^2 - n_3^2] \qquad (5.22)$$

Note that for each value of the index p, which describes the modal variation in the xz plane, there will be q solutions, $q = 0, 1, \ldots,$ describing the number of half-cycle variations along y within the core. Therefore, the parameter b' and the corresponding effective refractive index n'_{eff} must be designated by the pair of subscripts, p and q. The effective index $(n'_{\text{eff}})_{pq}$ can be expressed in terms of b_p and b'_{pq} by using Eqs. 5.19 and 5.22 as

$$(n'_{\text{eff}})_{pq} = [n_3^2 + b_p b'_{pq} (n_2^2 - n_3^2)]^{1/2} \qquad (5.23)$$

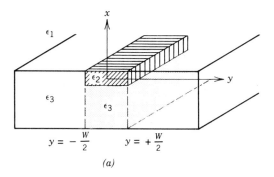

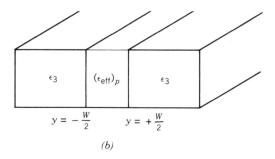

Figure 5.11 (a) Embedded-strip waveguide; (b) "equivalent" slab waveguide profile.

For the usual situation in which the difference in refractive index Δn between n_2 and n_3 is small then

$$n_2^2 - n_3^2 \cong 2 \, \Delta n \, n_3$$

and we have

$$(n'_{\text{eff}})_{pq} \cong n_3 + \Delta n \, b_p b'_{pq}$$

The associated propagation constant for each mode, $(k_z)_{pq}$, is therefore given by

$$(k_z)_{pq} = k_0 \, (n'_{\text{eff}})_{pq}$$

Let the corresponding transverse wavenumbers in the $\hat{y}$ direction for the regions $|y| < W/2$ and $|y| > W/2$ be defined by k_y and α_y, respectively. These can now be obtained by using the appropriate dispersion relation in each respective region and are given by

$$(\alpha_y)_{pq} = \sqrt{(k_z)_{pq}^2 - k_3^2} = k_0 \, \sqrt{(n'_{\text{eff}})_{pq}^2 - n_3^2}$$
$$\cong k_0 \, (2 \, \Delta n \, n_3 b_p b'_{pq})^{1/2}$$

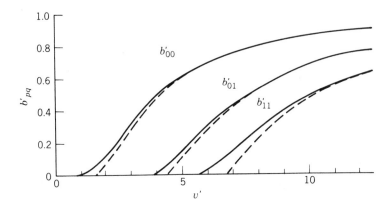

Figure 5.12 Comparison of $b'-v'$ diagram using effective index method (solid lines) with computer solution (dashed lines) (Ref. 3). For the above calculations it was assumed that $d = W$ and $\epsilon_1 = \epsilon_3$.

and

$$(k_y)_{pq} = \sqrt{k_0^2 (n_{\text{eff}})_p^2 - (k_z)_{pq}^2} = k_0 \sqrt{(n_{\text{eff}})_p^2 - (n'_{\text{eff}})_{pq}^2}$$
$$\cong k_0 [2 \, \Delta n \, n_3 \, b_p \, (1 - b'_{pq})]^{1/2}$$

A comparison of the accuracy of the technique described above with "exact" computer solutions is shown in Figure 5.12. Note that as expected, when the modes are far from cutoff (and thus propagate most nearly along z), the approximate solutions are in excellent agreement.

Several modes along with field and intensity plots are shown in Figure 5.13.

Example 5.2

Compute the effective refractive index for the fundamental $\hat{y}$-polarized mode on an embedded-strip 3-D waveguide having the following parameters:

$$n_1 = 1.0 \qquad d = 1.5\lambda_0$$
$$n_2 = 2.234 \qquad W = 2.5\lambda_0$$
$$n_3 = 2.214$$

Solution

Because the mode is $\hat{y}$ polarized, it is TE with respect to the xz plane. In this plane we have

$$a^{\text{TE}} = (n_3^2 - n_1^2)/(n_2^2 - n_3^2) = 43.9$$
$$v = k_0 d(n_2^2 - n_3^2)^{1/2} = 2.81$$

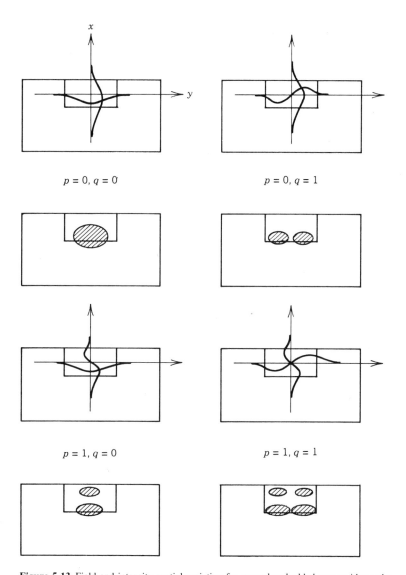

Figure 5.13 Field and intensity spatial variation for several embedded waveguide modes.

From Figure 4.11, using the normalized curve for $a \cong 40$ we obtain

$$b_0 = 0.44 \qquad b_p = 0 \qquad p > 0$$

so that from Eq. 5.19

$$(n_{\text{eff}})_0 = [(n_2^2 - n_3^2)b_0 + n_3^2]^{1/2} = 2.223$$

Next we compute the yz plane. From Eq. 5.21,

$$v' = k_0 W[(n_{\text{eff}})_0^2 - n_3^2]^{1/2} = 3.11$$

Note that $a' = 0$ since the guide is symmetric. Although the mode appears $\underline{\text{TM}}$ with respect to the yz plane, since the difference between (n_{eff}) and n_3 is small, we can use the normalized dispersion curve for TE modes. Thus we obtain

$$b_{00}' = 0.64 \qquad b_{pq}' = 0 \quad p \neq q \neq 0$$

and from Eq. 5.23,

$$(n_{\text{eff}})_{00}' \cong n_3 + (n_2 - n_3)b_0 b_{00}' = 2.220$$

5.3.2   Extension to Other Waveguide Geometries

The technique described above is readily extended to analyze the other waveguide geometries discussed in the Introduction. Consider, for example, the rib guide shown in Figure 5.2b. Viewed in the xz plane, the guide is observed to be composed of two different asymmetric slab structures, one for $|y| > W/2$ and one for $|y| \leq W/2$. These are shown in Figures 5.14a and 5.14b. For each of the two cross sections, an effective permittivity can be computed. Let us, therefore, define ϵ_{effa} and ϵ_{effb} as the effective permittivities computed in the xz plane for the asymmetric slabs shown in Figures 5.14a and 5.14b, respectively. These effective permittivities can be computed from the normalized dispersion curve in the same fashion described previously. Their values are then used to replace the true ridge waveguide with the equivalent symmetric structure shown in Figure 5.14c, thereby allowing an overall effective permittivity to be cal-

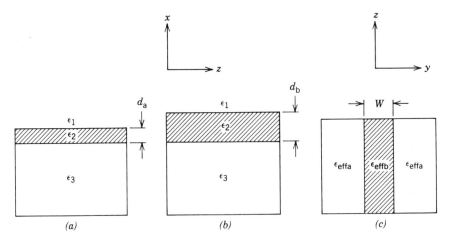

Figure 5.14 Rib waveguide. (a) $|y| > W/2$; (b) $|y| < W/2$; (c) "equivalent" symmetric structure.

culated. A similar procedure can be used to analyze the strip-loaded waveguide shown in Figure 5.2c.

5.4 RADIATION BENDING LOSSES

When a number of optical components are integrated onto a single chip, the dielectric waveguide pathways required to interconnect them often contain bends as shown in Figure 5.4. As we describe in this section, changes in the propagation direction for confined radiation invariably lead to a portion of the power being radiated away from the guide.

To analyze the effect of a bend, consider the dielectric waveguide having a curved section of radius R shown in Figure 5.15. The propagation direction z is defined to be along an axis of the waveguide. The geometry shown may be considered to correspond to the side view of a section of the symmetric slab waveguide discussed in Chapter 4. More practically, it could represent the top view of a three-dimensional waveguide such as that described in Section 5.3, with ϵ_1 and ϵ_2 signifying the waveguide effective dielectric constants.

Let us consider the behavior of the guided wave as it enters the bend. For simplicity, it is assumed that only the fundamental mode is above the cutoff frequency. If the radius of curvature is large, then intuitively it is expected that the properties of the mode for any point z along the bend should not differ significantly from the straight

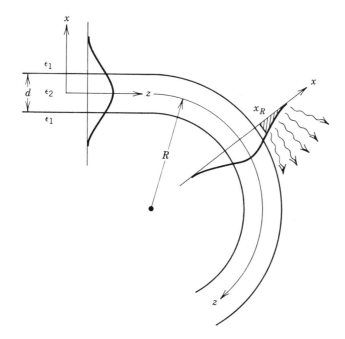

Figure 5.15 Section of a curved slab waveguide.

waveguide limit ($R \rightarrow \infty$). Constant phase fronts are therefore perpendicular to z, that is, in a radial direction, and the field distribution measured along such a phase front should closely resemble that of the mode in the straight guide limit. Further, the phase velocity $v_p = \omega/k_z$ measured along the axis of the bend should approximate that for $R \rightarrow \infty$. This must be equal to the tangential velocity at a distance R from the center of curvature or

$$\frac{\omega}{k_z} = R \frac{d\theta}{dt} \tag{5.24}$$

Consider now the tangential velocity of points on the constant phase front which are farther from the center of curvature. Since these points lie on the same phase front they all have the same angular velocity $d\theta/dt$. Their tangential velocity $v_{tan} = (x + R) d\theta/dt$ therefore increases with x and at some point x_r this velocity will exceed the velocity of light in the surrounding medium, region 1. Mathematically, this transition point is defined by the relation

$$\frac{\omega}{k_1} = (x_r + R) \frac{d\theta}{dt} \tag{5.25}$$

where

$$k_1 = \omega \sqrt{\mu \epsilon_1}$$

By combining Eq. 5.24 with 5.25, x_r is given in terms of the waveguide parameters by

$$x_r = \left(\frac{k_z - k_1}{k_1} \right) R$$

The physical significance of x_r can be interpreted with reference to Figure 5.15. As the guided mode enters the bend, the portion of the evanescent "tail" beyond x_r cannot travel sufficiently quickly to stay in phase with the remainder of the guided mode and therefore must split away from the guide and radiate into region 1. It is important to understand this does not imply that once the power contained in the "tail" at $z = 0$ is radiated away no tail exists. Rather, the tail is continually resupplied with power from the remainder of the guided mode as it travels along the bend. As a consequence, however, the overall amplitude of the mode must decrease with increasing propagation distance.

Let $P(z)$ be the total power contained in the mode at any point z along the bend. If we assume that the rate of loss of total power contained in the waveguide at z, $-dP(z)/dz$, is proportional to the power contained in the mode at that point, then

$$\frac{-dP(z)}{dz} = \alpha P(z) \tag{5.26}$$

where α is the proportionality constant. The above equation has the solution

$$P(z) = P(0)e^{-\alpha z} \tag{5.27}$$

and α is identified as the exponential decay constant.

To compute the attenuation, Eq. 5.26 is solved for α yielding

$$\alpha = \frac{-dP(z)/dz}{P(z)} \tag{5.28}$$

If the amount of power lost between the arbitrary planes z and $z + \Delta z$ is defined as $P_r(z)$, then for small Δz, α is approximated as

$$\alpha \cong \frac{P_r(z)}{\Delta z \, P(z)} \tag{5.29}$$

The quantity $P_r(z)/\Delta z$ can be obtained approximately by calculating the power in the evanescent tail along some plane z and estimating the distance Δz required for all the power contained in the tail along this plane to be radiated away. That is,

$$\frac{P_r}{\Delta z} = \frac{\text{power contained in tail at } z}{\text{distance for power in tail to radiate away}} \tag{5.30}$$

The distance Δz required for the tail to radiate away can be estimated by treating that portion of the waveguide field beyond $x = x_r$ as an equivalent radiating aperture as shown in Figure 5.16. The distance Δz may then be viewed as the diffraction distance and can be computed in a fashion analogous to that used for the uniform aperture discussed in Section 2.9. As was shown, if the field distribution of an aperture at some arbitrary plane, say $z = 0$, is given by $\psi(x, z = 0)$ then for any $z > 0$, ψ

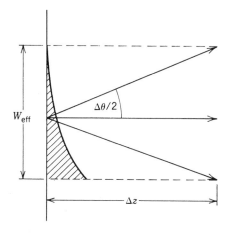

Figure 5.16 Equivalent radiating aperture for fields on a curved waveguide.

can be represented by a superposition of plane waves as

$$\psi(x, z) = \int_{-\infty}^{\infty} \psi(k_x) e^{-jk_x x} e^{-jk_z z} \, dk_x$$

where $\psi(k_x)$ is the Fourier transform of $\psi(x, z = 0)$. For the present case, the radiating aperture is

$$\psi(x, z = 0) = \psi(x_r) e^{-\alpha_x(x - x_r)} \qquad x \geq x_r$$

$$= 0 \qquad x < x_r$$

where α_x is the guided-mode transverse decay coefficient. Taking the Fourier transform we find straightforwardly that

$$|\psi(k_x)| = \frac{|\psi(x_r)|}{2\pi \, (\alpha_x^2 + k_x^2)^{1/2}} \tag{5.31}$$

Because $|\psi(k_x)|$ represents the amplitude of the plane waves making up the radiating aperture, we see from Eq. 5.31 that only those waves for which k_x is roughly less than or equal to α_x contribute significantly to the radiated field. In region 1 where the radiation occurs, k_x is related to k_1 via the dispersion relation

$$k_x^2 + k_z^2 = k_1^2$$

Provided that $k_x \ll k_1$ the corresponding angular spread, $\Delta\theta$, of plane waves with respect to the propagation direction is given by

$$\tan \frac{\Delta\theta}{2} \cong \frac{\Delta\theta}{2} = \frac{k_x}{k_1} \cong \frac{\alpha_x \lambda_1}{2\pi}$$

As with the uniform aperture of Section 2.9, Δz is defined as that distance required for the radiated plane waves to distribute their intensity over a transverse distance equal to that of the aperture. Since the field at the aperture varies as $e^{-\alpha_x x}$, the aperture width can be approximated as $W_{\text{eff}} \cong 1/\alpha_x$. Thus from Figure 5.16 we have that

$$\frac{W_{\text{eff}}}{\Delta z} \cong \Delta\theta = \frac{\alpha_x \lambda_1}{\pi}$$

and therefore

$$\Delta z \cong \frac{\pi W_{\text{eff}}}{\alpha_x \lambda_1} = \frac{\pi}{\alpha_x^2 \lambda_1} \tag{5.32}$$

which is the desired result.

To compute $P_r(z)$ we need simply to integrate the time-average Poynting power in the z direction from $x = x_r$ to infinity. Thus, the power radiated away per unit length along y is

$$P_r = \tfrac{1}{2} \operatorname{Re} \int_{x_r}^{\infty} (\mathbf{E} \times \mathbf{H}^*) \cdot \hat{z} \, dx \tag{5.33}$$

where the fields, $\mathbf{E}$ and $\mathbf{H,}$ are approximated as those of the guided mode in the straight waveguide limit. Similarly, the total power contained in the guide is found by integrating the time-average Poynting power from $x = -\infty$ to $x = +\infty$.

$$P = \tfrac{1}{2} \operatorname{Re} \int_{-\infty}^{\infty} (\mathbf{E} \times \mathbf{H}^*) \cdot \hat{z} \, dx \tag{5.34}$$

Making use of Eqs. 5.32–5.34 and 5.29 allows the attenuation coefficient to be computed.

Let us calculate α for the fundamental waveguide mode on the symmetric slab, the TE_0 mode. In Section 4.3.5, it was shown that for TE modes

$$\tfrac{1}{2} \operatorname{Re}(\mathbf{E} \times \mathbf{H}^*) \cdot \hat{z} = -\tfrac{1}{2} \operatorname{Re} E_y H_x^*$$

$$= \frac{k_z}{2\omega\mu} |E_y|^2$$

Using the expression for E_y given by Eq. 4.5 we have

$$P_r = \frac{k_z}{2\omega\mu} \int_{x_r}^{\infty} |E_y|^2 \, dx$$

$$= \frac{k_z}{2\omega\mu} |A|^2 \cos^2(k_{2x}d/2)e^{\alpha_x d} \int_{x_r}^{\infty} e^{-2\alpha_x x} \, dx \tag{5.35}$$

$$= \frac{k_z}{4\omega\mu\alpha_x} |A|^2 \cos^2(k_{2x}d/2)e^{\alpha_x(d-2x_r)}$$

The total power P is similarly obtained from Eq. 5.34 and was evaluated in Section 4.3.5 as

$$P = \frac{k_z}{4\omega\mu} |A|^2 d_{\text{eff}} \tag{5.36}$$

where

$$d_{\text{eff}} = (d + 2/\alpha_x)$$

Making use of Eqs. 5.29, 5.32, 5.35, and 5.36, we have for our estimate of attenuation the expression

$$\alpha = \frac{\alpha_x^2 \lambda_1}{\pi(\alpha_x d + 2)} \cos^2(k_{2x}d/2)e^{\alpha_x d} \exp[-2\alpha_x(k_z - k_1)R/k_1] \qquad (5.37)$$

The above results shows that the attenuation is of the functional form

$$\alpha = C_1 e^{-C_2 R}$$

where C_1 and C_2 are independent of the radius R. It is important to note that the radiation loss increases exponentially with decreasing bending radius. For $C_2 R \gg 1$, the attenuation is negligible, but rapidly becomes significant as $C_2 R$ approaches unity.

To evaluate the explicit dependence of the attenuation on the various waveguide parameters let us examine Eq. 5.37 in the usual limit where the fractional difference between core and surrounding refractive indices is small. Under these conditions only the fundamental mode is below cutoff and is weakly guided. For weak guidance, the bounce angle inside the core approaches 90° and the evanescent fields therefore extend far outside of the core region so that the field decay constant α_x is small. In this limit the propagation constant k_z can be approximated as follows:

$$k_z = (k_1^2 + \alpha_x^2)^{1/2} = k_1(1 + \alpha_x^2/k_1^2)^{1/2}$$

$$\cong k_1 + \frac{1}{2}\frac{\alpha_x^2}{k_1}$$

Substituting the above approximate expression for k_z into Eq. 5.37 we obtain for the constants C_1 and C_2

$$C_1 = \alpha_x^2/k_1$$

$$C_2 = \alpha_x^3/k_1^2$$

where it has been assumed that $\alpha_x d \ll 1$.

A more mathematically rigorous but significantly more complicated analysis (Ref. 7) shows that the constant C_2 should be replaced with C_2' where

$$C_2' = \tfrac{2}{3}C_2$$

so that

$$\alpha = \alpha_x^2/k_1 \exp[-\tfrac{2}{3}\alpha_x^3 R/k_1^2] \qquad (5.38)$$

The quantity α_x can be expressed in terms of the waveguide parameters (see problems) as

$$\alpha_x d = \sqrt{v^2 + 1} - 1 \qquad (5.39)$$

where the normalized film thickness v was defined in Section 4.3.4 as

$$v^2 = k_1^2\, d^2 \left(\frac{n_2^2}{n_1^2} - 1 \right)$$

Since the values of the refractive indices in regions 1 and 2 are generally very close to each other, v^2 can be accurately approximated as

$$v^2 \cong 2k_1^2\, d^2 \left(\frac{\Delta n}{n} \right) \tag{5.40}$$

where

$$\frac{\Delta n}{n} = \frac{n_2 - n_1}{n_1} \ll 1$$

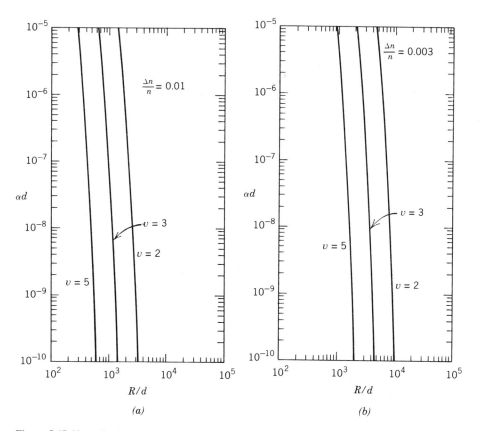

Figure 5.17 Normalized attenuation versus normalized bending radius for several values of parameter v. (a) $\Delta n/n = 0.01$; (b) $\Delta n/n = 0.003$.

Making use of Eqs. 5.39 and 5.40 in 5.38 allows us to obtain a simple expression for the normalized attenuation coefficient αd in terms of only three simple parameters, v, $\Delta n/n$, and R/d. The dependence of normalized attenuation upon normalized bending radius based upon these simplifying assumptions is illustrated in Figure 5.17 for three values of v and two values of $\Delta n/n$. These results are in good agreement with more rigorous but significantly more complicated derivations published elsewhere (Refs. 6, 7). Note in particular, the extremely rapid change in the attenuation coefficient with change in bending radius.

PROBLEMS

5.1 Give an expression for the cutoff frequency f_{cp} for the pth mode on a waveguide having the following refractive index profile:

$$n(x) = n_{max} [1 - (x/x_0)^2]^{1/2} \qquad |x| < x_A$$

$$n(x) = n_{min} \qquad |x| > x_A$$

and

$$n_{min} < n_{max}$$

5.2 Using the WKB method write down, but do not solve, the guidance condition for the waveguiding structures shown in Figure 5.18. What is the range of $(n_{eff})_p$ for guided modes in each case?

5.3 Compute the cutoff frequencies f_{cp} for each of the waveguides shown in problem 5.2 in terms of the variables n_1–n_4, c, and p.

5.4 A dielectric waveguide has a variation in refractive index which may be approximated as

$$n^2(x) = n_{max}^2 [1 - (x/x_0)^2] \quad ,$$

where

$$n_{max} = 1.5$$
$$x_0 = 11.6\lambda_0$$

(a) Compute $(n_{eff})_p$ and the turning points, $(x_{A,B})_p$, for the first three modes.
(b) What is the maximum angle θ_p that the ray associated with each mode makes with the z axis?

5.5 Write a computer program or solve Eq. 5.7 iteratively using a calculator to compute the effective index $(n_{eff})_p$ by WKB method for the first two modes for the following Gaussian permittivity profile waveguide:

$$\epsilon(x) = a\epsilon_0 \exp[-x^2/(b\lambda_0^2)]$$

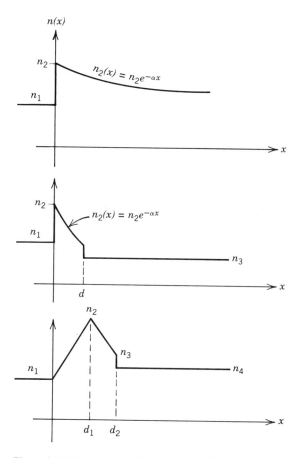

Figure 5.18 Various waveguide geometries treated by the WKB method.

where

$$a = 2.25$$
$$b = 3$$

5.6 Show that for the paraxial ray trajectory given in Eq. 5.17 to be self-consistent it is required that

$$\left| \frac{(2p + 1)}{k_0 \, n_{max} \, x_0} \right| \ll 1$$

5.7 Using the results of problem 5.6, show that for paraxial modes in a parabolic guide the group velocity $d\omega/dk_z$ is to first order independent of the mode number p. Explain your answer physically. (*Hint.* Consider the ray path followed by two different modes. Examine the respective path length and velocity along the path of each mode over one cycle.)

5.8   An optical waveguide is fabricated having a linear variation in permittivity given by

$$\epsilon(x) = \epsilon(1 - \alpha|x|) \qquad |x| < x_0$$

$$= \epsilon/10 \qquad\qquad |x| > x_0$$

(a) Use WKB theory to show that the propagation constant for the pth mode, $(k_z)_p$, is given by

$$(k_z)_p = k\sqrt{1 - [(p + \tfrac{1}{2}) \, 3\pi\alpha/4k]^{2/3}} \qquad k = \omega\sqrt{\mu\epsilon}$$

(b) What is the constraint placed upon p such that the modes can be considered well confined?

5.9   A ridge waveguide has the geometry shown in Figure 5.2b. Looking down from the top, the guide can be viewed as a symmetric structure with effective dielectric constants ϵ_a and ϵ_b, as shown in Figure 5.14c.

(a) Assuming TE polarization, compute ϵ_a and ϵ_b using Figure 4.11a and the following data:

$$
\begin{aligned}
\lambda_0 &= 0.8 \ \mu m & n_3 &= 2.214 \\
n_1 &= 1 & d_a &= 1 \ \mu m \\
n_2 &= 2.234 & d_b &= 1.8 \ \mu m
\end{aligned}
$$

(b) Compute $(n'_{eff})_{pq}$ for each propagating mode if $W = d_a$.

5.10   The embedded-strip guide, shown in Figure 5.2a, has the following parameters:

$$
\begin{aligned}
n_1 &= 1 & d &= 1.5\lambda_0 \\
n_2 &= 2.2 & W &= d \\
n_3 &= 2.12
\end{aligned}
$$

(a) Compute $(n_{eff})_{pq}$ for each allowed mode.

(b) Make a rough sketch of the field intensity for each mode as would be observed in the xy plane.

5.11   It is desired to transfer power between a solid-state laser (component A) and a detector (component B) using the 3-D waveguide shown in Figure 5.19. The guide is an embedded strip having an effective core index for the fundamental mode n_{eff} surrounded by a substrate of index n. The waveguide width is d.

(a) If $d = \lambda_0 = 0.63 \ \mu m$, $n_{eff} = 1.5$, and $n = 1.48$, compute the minimum area that the transition region may occupy if the radiation bending power loss through the curved sections is required to be less than 1%.

(b) Repeat (a) when $n_{eff} = 2.0$.

5.12   Derive Eq. 5.23.

5.13   Prove Eq. 5.39. *Hint.* Use the guidance condition given in Eq. 4.6 noting that near cutoff, $k_{2x}d$ is small for the fundamental mode.

5.14   (a) Make a plot of radiation loss in decibles (dB) per meter (m) versus $\Delta n/n$

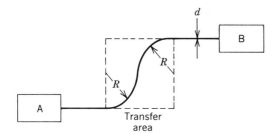

Figure 5.19 Interconnection between two integrated optics components using a curved section of waveguide.

for the fundamental mode propagating on a slab guide having $R = 1$ m, $d = 3$ μm, and $\lambda = 1$ μm.

(b) On the same graph make a plot of loss versus $\Delta n/n$ for $R = 2$ cm.

Note. The power in decibles is given by $P_{dB}(z) = 10 \log(e^{-\alpha z})$.

REFERENCES

1. Arnaud, J. A. "Transverse coupling in fiber optics, Part III: Bending losses." *Bell System Technical Journal* 53 (1974): 1379–1395.

2. Barnoski, M. K. *Introduction to Integrated Optics.* New York: Plenum, 1974.

3. Hocker, G. B., and Burns, W. K. "Mode dispersion in diffused channel waveguides by the effective index method." *Applied Optics* 16 (1977): 113–118.

4. Hocker, G. B., and Burns, W. K. "Modes in diffused optical waveguides of arbitrary index profile." *IEEE Journal of Quantum Electronics* QE-11 (1975): 270–276.

5. Hunsperger, R. G. *Integrated Optics: Theory and Technology.* New York: Springer-Verlag, 1982.

6. Marcatili, E. A. J. "Bends in optical dielectric guides." *Bell System Technical Journal* 48 (1969): 2103–2133.

7. Marcuse, D. *Light Transmission Optics.* New York: Van Nostrand-Reinhold, 1972.

8. Miller, S. E. "Directional control in light-wave guidance." *Bell System Technical Journal* 43 (1964): 1727–1738.

9. Snyder, A. W., White, I., and Mitchell, D. J. "Radiation from bent optical waveguides." *Electronics Letters* 11 (1975): 332–333.

10. Suematsu, Y., and Iga, K. *Introduction to Optical Fiber Communications.* New York: Wiley, 1982.

11. Tamir, T., ed. *Integrated Optics.* New York: Springer-Verlag, 1975.

12. Tyras, G. *Radiation and Propagation of Electromagnetic Waves.* New York: Academic, 1969.

CHAPTER 6

The Prism Coupler

6.1 INTRODUCTION

One of the primary diagnostic tools for evaluating the propagation characteristics of dielectric waveguides is the prism coupler. In its simplest form, the coupler makes use of a high-refractive-index prism placed in close proximity to a slab dielectric waveguide as shown in Figure 6.1. When an optical beam passing through the prism is incident upon its bottom at an angle exceeding the critical angle, the evanescent fields that extend below the prism base penetrate into the waveguide. These fields, as we show, are capable of transferring power between the incident beam and a waveguide mode. By appropriate choice of the angle of incidence and proper coupler design, a significant portion of the power in the incident beam may be transferred into a single chosen waveguide mode. Further, by reciprocity, if a second identical prism is placed in close proximity to the waveguide at some distance away from the input prism, each propagating mode will be coupled out of the guide at an angle that is characteristic of that particular mode. By measuring the output angles, a determination can be made of waveguide film refractive index and thickness.

To analyze the operation of the prism coupler, it is first necessary to understand why it is possible for power to be transmitted through the gap between the prism and waveguide even though the incident beam exceeds the critical angle at this prism–air interface. This phenomenon is referred to as frustrated total internal reflection.

6.2 FRUSTRATED TOTAL INTERNAL REFLECTION

The problem of transmission of a light beam through an air gap is a special case of plane-wave transmission through two dielectric interfaces as shown in Figure 6.2. For the prism coupler configuration, regions 4, 1, and 2 correspond, respectively, to the prism, air gap, and waveguide film.

Let us assume initially that a plane wave is incident from region 4 at an angle θ_i such that the critical angle is not exceeded at either the upper or lower interface. This restriction will be subsequently removed. To calculate the transmission coefficient, use is made of the so-called ray summation technique. The incident and transmitted plane waves are considered to be made up of an infinite number of pencil beams or rays, a few of which are shown in Figure 6.3. Depending on whether the incident wave is TE or TM, the amplitude of each ray is chosen to be proportional to the electric or magnetic field strength, respectively. For convenience, the amplitude of the rays corresponding

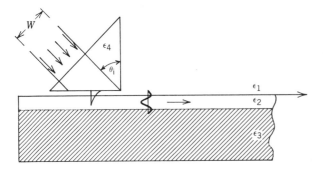

Figure 6.1 Geometry for prism coupler.

to the incident field are normalized to unity. What we wish to determine is how each incident ray will contribute to any arbitrary single transmitted ray. Let us, therefore, examine the contributions to the transmitted ray T_d. We number the incident rays consecutively 1, 2, 3 . . . . The phase of each incident ray is referenced relative to any arbitrary constant phase plane such as the one indicated by the dashed line in Figure 6.3. From this figure, it is observed that ray 1 contributes directly to the transmitted ray T_d by propagating along path 1–1″, being partially transmitted and reflected at the upper and lower dielectric interfaces. Let the phase shift resulting from propagation along 1–1″ be given by ϕ_0. Further, let the respective transmission coefficients associated with upper and lower interfaces be defined by T_{41} and T_{12}. These are obtained by direct evaluation of the TE or TM wave transmission coefficients for a single dielectric interface derived in Chapter 3. Recalling that each incident ray has been normalized to unity, the contribution from ray 1 to output ray T_d is then given by

$$T_{41}T_{12}e^{j\phi_0} \tag{6.1}$$

To obtain the contribution to the output from ray 2 we note from Figure 6.3 that this ray does not contribute to T_d along a direct path. Instead, the portion of the ray which is transmitted through the upper interface travels along the zigzag path 2′–2″–

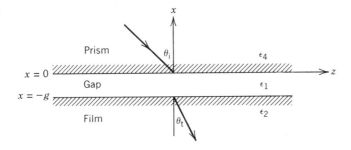

Figure 6.2 Transmission of a light beam through an air gap.

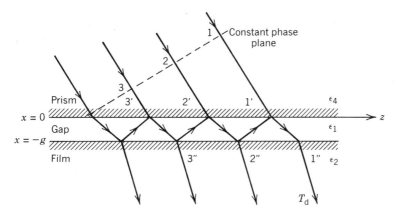

Figure 6.3 Contribution of several incident rays to the transmitted ray T_d.

$1'$–$1''$. Along this path the ray is partially reflected at the lower and upper dielectric boundaries prior to being partially transmitted through the lower interface where it contributes to ray T_d. Let the reflection coefficient seen by ray 2 at the upper and lower boundaries be denoted by R_{14} and R_{12}, respectively, and let the additional phase shift encountered by ray 2 relative to ray 1 due to a longer propagation path be given by $\Delta\phi$. Then the contribution for ray 2 to the output ray T_d is equal to that from ray 1 multiplied by the additional factor

$$R_{14}R_{12}e^{j\Delta\phi} \qquad (6.2)$$

To compute the phase shift $\Delta\phi$ we make reference to Figure 6.4. The indicated dashed lines perpendicular to the pencil beams represent arbitrary constant phase planes in regions 4 and 1. Thus, the phase shift due to propagation of ray 2 along the path between A and A' must be equal to that for ray 1 between C and C'. The net phase difference $\Delta\phi$ therefore results from ray 2 traversing the additional path length $l_{A'BC'}$ relative to that of ray 1. The phase shift along this path was computed in Section 4.3.3 where it was shown that $\Delta\phi$ is given by

$$\Delta\phi = -2k_{1x}g \qquad (6.3)$$

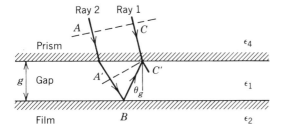

Figure 6.4 Arbitrary constant phase planes (dashed lines) in the prism and air gap.

By symmetry, $\Delta\phi$ must also be the relative phase shift between any two adjacent incident rays. Thus the contribution to the output pencil beam T_d from each subsequent ray 3, 4, . . . is obtained by multiplying the previous ray's contribution by the factor given in expression 6.2. The total contribution to the transmitted ray T_d is therefore

$$T_d = T_{41}T_{12}e^{j\phi_0} \sum_{n=0}^{\infty} (R_{14}R_{12}e^{j\Delta\phi})^n \tag{6.4}$$

The above expression is an infinite geometric sum and may be cast in the form

$$T_d = T_{41}T_{12}e^{j\phi_0} \lim_{N\to\infty} S_N$$

where

$$S_N \equiv \sum_{n=0}^{N-1} a^n$$

and

$$a = R_{14}R_{12}e^{j\Delta\phi}$$

The sum S_N is readily solved in closed form by noting that

$$S_{N+1} = S_N a + 1$$

and

$$S_{N+1} - S_N = a^N$$

which, eliminating S_{N+1}, yields

$$S_N = \frac{1 - a^N}{1 - a} \tag{6.5}$$

Provided that $|a| < 1$ then in the limit as $N \to \infty$, $a^N \to 0$ and

$$\lim_{N\to\infty} S_N = \frac{1}{1 - a}$$

We, therefore, obtain for the transmission coefficient for a plane wave propagating

downward through the gap

$$T_d = \frac{T_{41}T_{12}e^{j\phi_0}}{1 - R_{14}R_{12}e^{-j2k_{1x}g}} \qquad (6.6)$$

To further show explicitly the dependence of T_d on the gap thickness, we note that the phase shift ϕ_0 can be broken into two parts, a portion in region 4 along the path 1–1' and a portion in region 1 along 1'–1". The phase shift for the first portion is simply minus the product of the wavenumber k_4 with the segment length $l_{1'-1'}$, or

$$\phi_{1-1'} = -k_4 l_{1-1'}$$

To obtain the phase shift along the second segment reference is made to Figure 6.5. The geometry shows that the segment 1'–1" can be represented as the sum of two smaller paths

$$l_{1'-1''} = (g \cos \theta_g + \Delta z \sin \theta_g)$$

and, therefore, the corresponding phase shift is

$$\phi_{1'-1''} = -k_1(g \cos \theta_g + \Delta z \sin \theta_g)$$
$$= -k_{1x}g - k_4 \Delta z \sin \theta_i$$

where we have made use of Snell's law to relate k_4 to k_1. The total phase shift ϕ_0 is therefore given by

$$\phi_0 = \phi_{1-1'} + \phi_{1'-1''}$$
$$= -k_{1x}g - k_4 (l_{1-1'} + \Delta z \sin \theta_i)$$

Because the reference phase plane in region 4 may be chosen arbitrarily, the second term in the above expression may be made equal to a multiple of 2π by appropriate

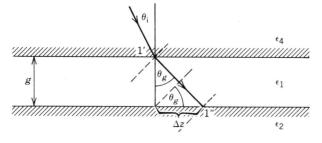

Figure 6.5 Geometrical construction for calculating the phase shift along the path segment $l_{1'-1''}$.

choice of the length $l_{1-1'}$. Thus, for such a suitably chosen reference plane $\phi_0 = -k_{1x}g$ and the transmission coefficient for ray T_d is given by

$$T_d = \frac{T_{41}T_{12}e^{-jk_{1x}g}}{1 - R_{14}R_{12}e^{-j2k_{1x}g}} \tag{6.7}$$

Note that although the transmission coefficient was obtained explicitly for only a single transmitted ray, the same result must apply to all transmitted rays because of symmetry.

Let us examine this expression now for the case in which the critical angle is exceeded at the upper interface but not at the lower. When the incident angle exceeds θ_c then we have shown that in the gap the fields are evanescent and the transverse wavenumber k_{1x} goes over to $-j\alpha_{1x}$. Thus, in this limit

$$T_d = \frac{T_{41}T_{12}e^{-\alpha_{1x}g}}{1 - R_{14}R_{12}e^{-2\alpha_{1x}g}} \qquad \theta_i > \theta_c \tag{6.8}$$

This expression, therefore, represents an effective transmission coefficient between half-space region 4 and half-space region 2 as shown in Figure 6.6. Based on the analysis for power transmitted through a single dielectric interface, we obtain for the power transmission coefficient through the gap

$$t_d^{TE} = \frac{k_{2x}}{k_{4x}} |T_d^{TE}|^2 \tag{6.9}$$

and

$$t_d^{TM} = \frac{\epsilon_4}{\epsilon_2} \frac{k_{2x}}{k_{4x}} |T_d^{TM}|^2 \tag{6.10}$$

These expressions imply that time-average power is now transmitted through the gap. The quantum mechanical equivalent of this phenomenon is known as "tunneling."

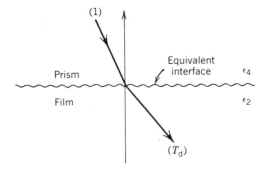

Figure 6.6 "Equivalent" interface between prism and film.

To appreciate these results, let us consider the situation in which the gap size g is chosen in such a fashion so that $|\alpha_{1x}g| \gg 1$. Then the magnitude of the transmission coefficient is small and is given to good approximation by

$$|T_d| \cong |T_{41}T_{12}e^{-\alpha_{1x}g}| \qquad (6.11)$$

This approximate solution is easily understood from the standpoint of the ray summation approach. We have shown that the transmitted ray is the superposition of an infinite number of bouncing pencil beams with each successive contribution making one additional pass downward and upward through the gap. When the critical angle is exceeded, these rays represent portions of an inhomogenous wave whose amplitude decays exponentially as it propagates across the gap. Thus, the contribution of each successive ray to the output ray T_d is exponentially smaller and, therefore, to good approximation only the contribution from ray 1 need be considered.

The magnitude of the contribution from ray 1 is simply the product of the magnitude of the individual interface transmission coefficients times the exponential decay factor $e^{-\alpha_{1x}g}$ representing the reduction in field amplitude resulting in passing through the gap. In terms of a wave-vector diagram we see that even though Snell's law cannot be represented graphically at each interface due to the evanescent nature of the fields in region 1, it may be drawn between region 4 and 2 as shown in Figure 6.7.

Under the weak transmission approximation, the form for the transmissivity can be shown to be particularly simple. Let us evaluate the expression in this limit for TE wave incidence. Making use of relation 6.9 and the approximate expression 6.11, the transmissivity is

$$t_d \cong \frac{k_{2x}}{k_{4x}}|T_{41}|^2|T_{12}|^2 e^{-2\alpha_{1x}g}$$

The single interface transmission coefficients for TE waves are obtained from Eq. 3.9

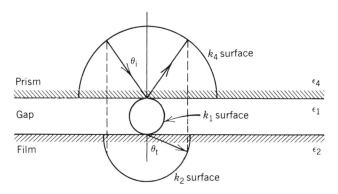

Figure 6.7 Wave surfaces in regions 1, 2, and 4. Note that phase matching between regions 4 and 2 is possible even though $k_1 < k_4$.

as

$$T_{41} = \frac{2}{1 - j\alpha_{1x}/k_{4x}}$$

$$T_{12} = \frac{2}{1 + jk_{2x}/\alpha_{1x}}$$

If we define

$$\tan(\phi_2/2) = \alpha_{1x}/k_{2x}$$

$$\tan(\phi_4/2) = \alpha_{1x}/k_{4x}$$

and note that $k_{2x}/k_{4x} = (k_{2x}/\alpha_{1x})/(k_{4x}/\alpha_{1x})$ then straightforward algebraic manipulation shows that

$$t_d = 4 \sin \phi_2 \sin \phi_4 \, e^{-2\alpha_{1x}g} \tag{6.12}$$

where

$$\sin(\phi_{2,4}/2) = \left[\frac{(\epsilon_4/\epsilon_1) \sin^2 \theta_i - 1}{(\epsilon_{2,4}/\epsilon_1) - 1} \right]^{1/2}$$

and

$$\alpha_{1x}g = \frac{2\pi g}{\lambda_1} (\epsilon_4/\epsilon_1 \sin^2 \theta_i - 1)^{1/2}$$

6.3   WAVEGUIDE EXCITATION USING THE PRISM COUPLER

The ray summation technique used to obtain the transmission coefficient through an air gap may also be applied to obtain the power coupled into a particular waveguide mode. As shown in Figure 6.8, a beam of radiation of width W is incident at an angle θ_i upon a prism coupler. The incident beam is again divided up into a number of rays or pencil beams. Each beam will be transmitted through the air gap to contribute in some manner to the output ray at the end of the coupler section. Whatever the resultant amplitude of this ray turns out to be, this amplitude will be maintained as the beam continues along its zigzag path in the $+z$ direction because no further mechanism exists for coupling out of or into the guide beyond the right edge of the prism coupler. The mechanism for buildup within the guide is readily analyzed by replacing the two interfaces between the prism base and top of the waveguide by a single effective boundary having the same transmission and reflection coefficients calculated in the last section. This is shown in Figure 6.8. The incident rays are again numbered 1, 2,

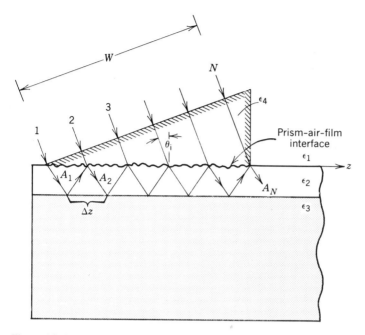

Figure 6.8 Contribution of N incident beam rays to the amplitude of guided-wave output ray A_N.

. . . , N. Note, however, that because of the finite width of the beam there are a finite number N of contributing rays. The amplitude of each corresponding transmitted ray in the waveguide is labeled A_n, $n = 1, 2, \ldots, N$. Thus, A_n represents the waveguide field amplitude under the prism at a point $z = (n - 1)\Delta z$ from the left edge of the prism and A_N represents the field amplitude transferred to the waveguide beyond the right end of the prism.

The contribution of each incident ray to A_n follows precisely the same analysis as was used in the last section. Let us define the reflection coefficient seen by a ray bouncing off the lower and upper waveguide surfaces by R_ℓ and R_u, respectively, and the transmission coefficient downward through the gap by T_d. The amplitude of A_n is then given by

$$A_n = T_d + T_d R_\ell R_u e^{-j2k_{2x}d} + \cdots + T_d (R_\ell R_u e^{-j2k_{2x}d})^{(n-1)}$$

$$= T_d \frac{1 - (R_\ell R_u e^{-j2k_{2x}d})^n}{1 - R_\ell R_u e^{-j2k_{2x}d}}$$

To evaluate the above expression we note first that for the wave to be guided by total internal reflection, the magnitude of the reflection coefficient of R_ℓ must be unity. Therefore,

$$R_\ell = e^{j\phi_\ell}$$

where ϕ_ℓ is the phase shift due to total internal reflection from the lower boundary as calculated in Chapter 3. The reflection at the upper boundary will be represented in terms of its magnitude and phase

$$R_u = |R_u|e^{j\phi_u}$$

Thus, A_n may be rewritten as

$$A_n = T_d \frac{1 - |R_u|^n e^{jn\phi}}{1 - |R_u|e^{j\phi}}$$

where

$$\phi = \phi_\ell + \phi_u - 2k_{2x}d \tag{6.13}$$

Note that ϕ represents the relative phase between successive ray contributions. The power reflection coefficient at the upper boundary, $r_u = |R_u|^2$, must by conservation of power be related to the power transmission coefficient upward through the gap by

$$r_u + t_u = 1$$

Therefore,

$$A_n = T_d \left[\frac{1 - (1 - t_u)^{n/2}e^{jn\phi}}{1 - (1 - t_u)^{1/2}e^{j\phi}} \right] \tag{6.14}$$

Let us adjust the prism coupling gap so that t_u is almost zero. This can always be accomplished since in the limit that g approaches infinity we have total internal reflection and t_u equals zero. In this limit the binomial approximation may be used to give

$$(1 - t_u)^{n/2} \cong 1 - \frac{n}{2}t_u$$

Further, provided also that $|(n/2)t_u| \ll 1$ for all n, then by Taylor series,

$$1 - \frac{n}{2}t_u \cong e^{-(n/2)t_u}$$

so that Eq. 6.14 may be rewritten as

$$A_n \cong T_d \left[\frac{1 - e^{-(n/2)t_u}\, e^{jn\phi}}{1 - (1 - t_u)^{1/2}e^{j\phi}} \right] \tag{6.15}$$

Additionally, let the angle of incidence θ_i be adjusted in such a manner that the phase term ϕ is close to a multiple of 2π. That is

$$\phi = 2p\pi + \delta$$

where

$$|\delta| \ll 1$$

Noting that

$$e^{j\phi} = e^{j(2p\pi + \delta)} = e^{j\delta} \cong 1 + j\delta$$

then algebraic manipulation of expression 6.15 yields

$$A_n = T_d \left[\frac{1 - e^{jn\delta}e^{-(n/2)t_u}}{t_u/2 - j\delta\,(1 - t_u/2)} \right]$$

Since by assumption both δ and t_u are small, the product term $\delta(t_u/2)$ in the above expression for A_n may be neglected and

$$|A_n| \cong A_\infty \left| 1 - e^{jn\delta}e^{-(n/2)t_u} \right| \tag{6.16}$$

where

$$A_\infty = \frac{|T_d|}{[(t_u/2)^2 + \delta^2]^{1/2}}$$

Let us investigate how A_n varies with distance along the prism. Note that from Figure 6.8, the distance z from the left end of the prism is given by $(n - 1)\Delta z$, so that increasing values of n correspond to increasing distance from the left end of the prism. We first examine the n-dependent term multiplying A_∞ by considering it to be the sum of two phasors. To the term $1e^{j0}$ we subtract a second term, $e^{jn\delta}e^{-(n/2)t_u}$ whose direction and magnitude depend on n and δ. Note that as n increases, the magnitude of the second phasor decreases. Thus, for various values of n the phasor sum R might look as shown in Figure 6.9a. The tip of the resultant vector versus increasing n would, therefore, look as shown in Figure 6.9b and $|A_n|$ would exhibit an oscillatory behavior as shown in Figure 6.10. Thus, when the phase difference between adjacent ray contributions is not exactly 2π the field amplitude builds up under the prism in an oscillatory manner with some rays adding constructively and some adding destructively. A_∞ is observed to be the saturation value for an infinitely wide beam. However, when the phase term $\delta = 0$, the oscillations cease, giving an exponential buildup with N (Figure 6.11). In this situation each additional pencil beam adds in phase with the one before it, asymptotically building up to the value A_∞ for large beamwidths.

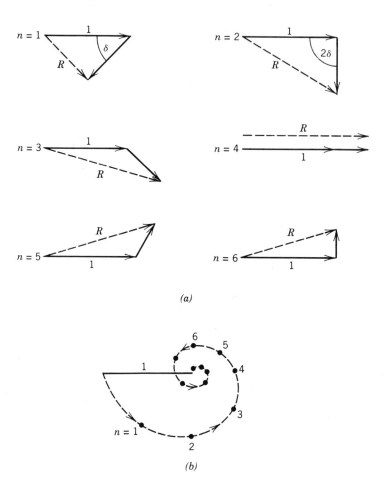

Figure 6.9 (*a*) Phasor representation of Eq. 6.16 for several values of $n = 1, 2, \ldots, N$. (*b*) Trace of the tip of the resultant vector R versus n.

Let us examine the dependence of A_∞ on the parameter δ. The ratio $|A_\infty (\delta \neq 0)|^2 / |A_\infty (\delta = 0)|^2$ is a measure of the relative power coupled into the waveguide and is given by

$$\left| \frac{A_\infty(\delta \neq 0)}{A_\infty(\delta = 0)} \right|^2 = \frac{1}{[1 + (2\delta/t_u)^2]} \tag{6.17}$$

which is plotted in Figure 6.12. Since the transmissivity through the coupler air gap may, in theory, be made as small as desired, the above curve can be very sharply peaked about $\delta = 0$ and δ can, therefore, be forced to be arbitrarily small for significant power buildup within the guide.

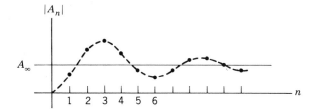

Figure 6.10 Buildup of guided-wave field amplitude $|A_n|$ with propagation distance under non-phase-matched conditions.

The fact that δ must be small for significant intensity buildup within the guide has a simple physical interpretation. Recall that the total phase difference between adjacent rays is given by

$$\phi = 2p\pi + \delta = \phi_l + \phi_u - 2k_{2x}d \qquad (6.18)$$

Suppose δ is not zero. If we sum up all N of these rays vectorially as shown in Figure 6.13, the net contribution to the output A_N tends to add up to zero whereas if they all add up in phase we get Figure 6.14. Note that under the condition $\delta = 0$ we obtain the requirement

$$\phi_l(\theta) + \phi_u(\theta) - 2k_{2x}(\theta)d = 2p\pi \qquad (6.19)$$

where θ is the bounce angle in the waveguide. This is precisely the guidance condition for the slab dielectric waveguide obtained in Chapter 4 with the exception that ϕ_u, the phase shift upon reflection from the top waveguide boundary in the presence of the prism coupler, replaces the upper interface phase shift for a dielectric waveguide in absence of the prism coupler. However, if the transmission coefficient of the coupler is small as we have assumed, ϕ_u approaches the value obtained in absence of the prism coupler. In this so-called weak coupling limit, for Eq. 6.19 to be satisfied, we require

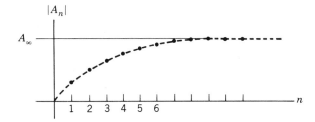

Figure 6.11 Buildup of guided-wave field amplitude with propagation distance under phase-matched conditions.

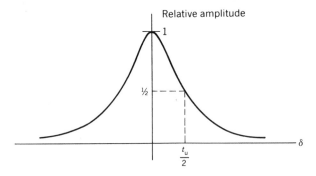

Figure 6.12 Ratio of coupled intensity under non-phase-matched conditions to that under phase-matched conditions as a function of mismatch parameter δ.

$\theta = \theta_p$ where θ_p corresponds to the bounce angle for the pth waveguide mode:

$$\tan \theta_p = (k_z)_p / (k_{2x})_p$$

The required angle of incidence θ_i for the incident beam inside the prism coupler is related to θ_p by Snell's law (phase matching)

$$k_4 \sin \theta_i = k_2 \sin \theta_p \qquad (6.20)$$

Thus, by adjusting θ_i properly, we can couple to any propagating mode desired. Relation 6.20 is an extremely important one. It states simply *that in order to couple to a particular waveguide mode we must use an incident beam which is phase matched to that mode.* That is,

$$(k_z)_{\text{incident}} = (k_z)_p \qquad (6.21)$$

Under this condition the incident and guided waves are said to be synchronous. The function of the prism now becomes clear. Notice that in absence of a prism having a permittivity greater than that of the waveguide film, the phase-matching criterion could not be satisfied by an incident plane wave. For example, a plane wave incident on the waveguide film in absence of the prism coupler would have a corresponding k diagram

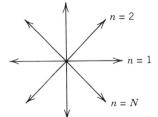

Figure 6.13 Effective cancellation of rays contributing to output when $\delta \neq 0$.

Figure 6.14 Summation of rays contributing to output when $\delta = 0$.

as shown in Figure 6.15. Because for guided waves $k_1 \leq k_{zp} < k_2$, then it is clear from Figure 6.15 that no angle of incidence can satisfy relation 6.21. However, if the prism permittivity ϵ_4 equals or exceeds that in the film we can have the situation shown in Figure 6.16 and phase matching is always possible.

To determine the coupling length necessary to transfer appreciable power into a particular mode of the waveguide recall that under phase-matched conditions ($\delta = 0$) the buildup of beam intensity goes as

$$A_n = A_\infty \left[1 - \exp\left(-\frac{n}{2} t_u \right) \right]$$

or

$$A(z_n) = A_\infty [1 - \exp(-z_n t_u / 2\Delta z)] \qquad (6.22)$$

where

$$z_n = n\Delta z$$

Thus, the distance required for the amplitude to build up to $1/e$ of its saturation value

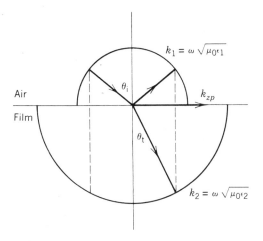

Figure 6.15 Without high-refractive-index prism coupler, phase matching of incident beam to guided wave is impossible.

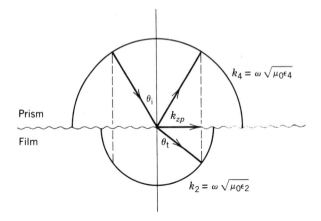

Figure 6.16 Introduction of prism permits phase matching to guided wave.

A_∞ is given by

$$z_n = \frac{2\Delta z}{t_u} \tag{6.23}$$

The parameter Δz can be related to the waveguide dimensions and mode propagation angle within the guide θ_p using Figure 6.17. Note that to properly relate the ray spacing to guide geometry in our ray approach we must incorporate the effect of the Goos–Hanchen shift discussed in Chapter 2, replacing the physical thickness d with the effective thickness d_{eff}. This dimension was shown in Section 3.6 to be

$$d_{\text{eff}} = d + d_1 + d_3 \tag{6.24}$$

where

$$d_{1,3} = \frac{1}{\alpha_{1,3x}} \qquad \text{for TE waves}$$

$$= \frac{q_{1,3}}{\alpha_{1,3x}} \qquad \text{for TM waves}$$

and

$$q_{1,3} = \left[\frac{k_{2x}^2 + \alpha_{1,3x}^2}{k_{2x}^2 + \left(\dfrac{\epsilon_2}{\epsilon_{1,3}}\right)^2 \alpha_{1,3x}^2} \right] \frac{\epsilon_2}{\epsilon_{1,3}}$$

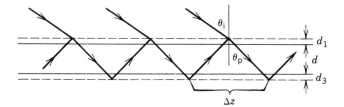

Figure 6.17 Proper computation of spacing Δz between adjacent rays requires use of "effective" waveguide boundaries resulting from Goos–Haenchen shift.

From the geometry then,

$$\Delta z = 2d_{\text{eff}} \tan \theta_p \qquad (6.25)$$

An explicit form for the transmissivity t_u for a wave traveling upward through the air gap is readily obtained from the transmissivity expression for downward propagation t_d. As is observed from comparison of Figures 6.18a and 6.18c, the two coefficients are measures of the power transmitted in opposite directions along the same ray path.

Intuitively we would expect that the amount of power transmitted across an interface should not depend upon the direction of that flow. To demonstrate this rigorously, we observe that the geometry of Figure 6.18a is obtainable from that of Figure 6.18c by the interchange of the roles of regions 2 and 4 as well as the incident and transmitted angles θ_i and θ_t. Correspondingly, the expression for t_u is, therefore, immediately derived from that for t_d given in Eq. 6.12 by interchange of the indices 2 and 4 as well as the angle θ_i and θ_t. If these interchanges are performed, it is easy to show that the result remains invariant. Thus, for example, we obtain for TE waves from expressions

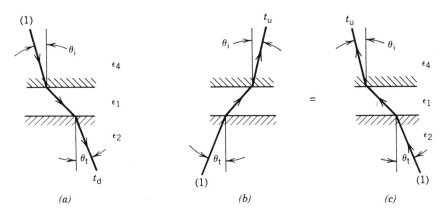

Figure 6.18 Geometry for computing transmissivity through the air gap. (a) Downward path for computing t_d; (b, c) equivalent upward paths for computing t_u.

6.12, 6.23, and 6.25 a $1/e$ buildup distance in the weak coupling limit given by

$$z_n = \frac{d_{\text{eff}} \tan \theta_p}{\sin \phi_2 \sin \phi_4} e^{2\alpha_{1x}g} \tag{6.26}$$

For many typical applications, coupling lengths are on the order of several thousand λ which for visible radiation corresponds to distances on the order of a few centimeters.

We have demonstrated that under phase-matched conditions the waveguide amplitude initially grows as the prism length L is increased. It might be hypothesized therefore, that by making L sufficiently large, all of the incident radiation could be coupled into the guide. This is, in fact, not the case since radiation is not only coupled into the waveguide through the air gap between prism and film but is also coupled out. When only a small percentage of the incident power has been transferred into the waveguide, the dominant direction of power flow is from the incident beam into the guiding film. However, when a sufficiently large percentage of incident power is coupled into the film the reverse process dominates. Steady state results when these two processes balance, leading, as we shall show, to a maximum theoretical coupling efficiency of about 80%.

We shall compute the maximum coupling efficiency using an extension of the ray technique. Let us assume TE incidence, as shown in Figure 6.19. The magnitude of the incident power density in the prism S_i is simply

$$S_i = |E|^2/\eta_4 = 1/\eta_4$$

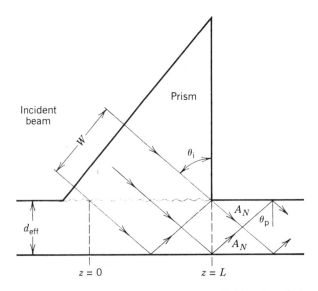

Figure 6.19 Geometry for calculating power coupled into the guided-wave mode.

where

$$\eta_4 = \sqrt{\mu_0/\epsilon_4}$$

and the electric field amplitude $|E|$ is normalized to unity for convenience. The total incident power P_i is equal to the product of S_i with the cross-sectional area of the incident beam W, or

$$P_i = S_i W = \frac{L \cos \theta_i}{\eta_4}$$

Consider next the total power flow from left to right across the waveguide entrance at the plane $z = L$. As shown in Figure 6.19, this power is contained in an upward and downward bundle of rays each having electric field amplitude A_N. The total power density incident on the entrance aperture from each bundle is

$$S_N = |A_N|^2/\eta_2 \qquad \eta_2 = \sqrt{\mu_0/\epsilon_2}$$

The total power P_N carried by these two bundles into the waveguide is equal to the product of the power density S_N and the waveguide aperture seen by the bundle, d_{eff} sin θ_p. Note that the effective waveguide thickness d_{eff} discussed in Section 3.6 is used rather than the true thickness d. Multiplying by two to account for the two bundles yields

$$P_N = 2 \frac{|A_N|^2}{\eta_2} d_{eff} \sin \theta_p$$

Making use of Eq. 6.16, P_N can be written under phase-matched conditions as

$$P_N = \frac{8 |T_d|^2}{\eta_2 t_u^2} [1 - e^{-(Lt_u/2\Delta z)}]^2 d_{eff} \sin \theta_p$$

Noting that $t_d = t_u$ and that

$$t_d = \frac{k_{2x}}{k_{4x}} |T_d|^2 = \frac{\eta_4 \cos \theta_p}{\eta_2 \cos \theta_i} |T_d|^2$$

we obtain for the coupling efficiency

$$\frac{P_N}{P_i} = \frac{2 [1 - e^{-(Lt_d/2\Delta z)}]^2}{(Lt_d/4d_{eff} \tan \theta_p)}$$

or

$$\frac{P_N}{P_i} = \frac{2\,(1 - e^{-SL})^2}{SL} \tag{6.27}$$

where

$$S = t_d/2\Delta z$$

To obtain the maximum power transfer, we differentiate Eq. 6.27 with respect to the normalized parameter SL and set the resulting expression equal to zero. This yields

$$(SL)_{max} \cong 1.257$$

Substituting the above value for $(SL)_{max}$ back into Eq. 6.27 yields as the maximum theoretical transfer efficiency

$$\left(\frac{P_N}{P_i}\right)_{max} \cong 81.4\%$$

It should be pointed out that by using a nonuniform gap the transfer efficiency can be increased to 100%.

6.4 FILM CHARACTERIZATION USING THE PRISM COUPLER

In the previous section, it was demonstrated that strong excitation of a guided mode requires very precise phase matching with the incident launching beam. The phase-matching requirement of Eq. 6.20 implies that for a waveguide mode p propagating with propagation constant $(k_z)_p$ there corresponds an angle for the incident beam, θ_i, which causes strong excitation. That is,

$$\frac{\omega}{c} n_4 \sin \theta_i = (k_z)_p = \frac{\omega}{c}(n_{eff})_p \tag{6.28}$$

By logical extension of the above analysis, we would expect that if a waveguide mode p is excited and a second prism placed in proximity with the guide, this prism will couple radiation out of the guide at the angle θ_i given by the above relation.

The fact that θ_i can be easily and precisely measured for each mode allows for a determination of the corresponding (n_{eff}).

Because $(n_{eff})_p$ depends on the refractive index n_2 and thickness d of the guiding film layer, experimental measurement of the effective index for a pair of modes provides a convenient method for determining these two parameters. Specifically, let us demonstrate how the refractive index n_2 and thickness d of the guiding film can be obtained.

Figure 6.20 shows the experimental setup for the measurement of the waveguide film properties. The waveguide containing the film under test is pressed against the base of a prism by means of a spring-loaded clamp. Dust particles between the prism and guide act as spacers producing the required air gap g. A second output prism is clamped in a similar fashion a short distance away along the propagation direction. In practice, a single prism serves for both input and output beams. The prism sits on a precision rotating stage that allows the angle of the incident beam to be varied with respect to the waveguide. Also mounted on the rotary table is an observation screen.

When the incident beam angle is adjusted to be synchronous with the fundamental mode, the guiding film appears bright, indicating strong excitation. Due to scattering from small waveguide irregularities, a portion of the power in the fundamental mode is coupled into a number of higher order modes. Each of the excited modes couples its power out of the output prism at an angle θ_i determined by Eq. 6.28 and therefore gives rise to a series of beams observed on the output screen, one for each mode.

Let us assume that the output angle θ_i is measured for the two lowest order modes, yielding from Eq. 6.28, $(n_{\text{eff}})_0$ and $(n_{\text{eff}})_1$. The effective indices are related to the film index n_2 and film thickness d via the guidance condition, Eq. 4.28, which can be written as

$$k_0 d[n_2^2 - (n_{\text{eff}})_p^2]^{1/2} = \psi_p[n_2, (n_{\text{eff}})_p] \qquad (6.29)$$

where

$$\psi_p = p\pi + \tfrac{1}{2}\phi_1[n_2, (n_{\text{eff}})_p] + \tfrac{1}{2}\phi_3[n_2, (n_{\text{eff}})_p]$$

Here ϕ_1 and ϕ_3 are the usual phase shifts due to total internal reflection at the upper and lower interface of the waveguide film. Their dependence on n_2 and $(n_{\text{eff}})_p$ has been explictly indicated. Substituting for $p = 0$ and $p = 1$ into Eq. 6.29 allows $k_0 d$ to be eliminated and results in a single equation for n_2^2 of the form

$$n_2^2 = F(n_2^2) \qquad (6.30)$$

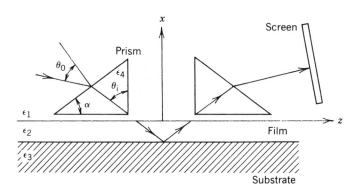

Figure 6.20 Schematic of setup for measuring waveguide film refractive index.

where

$$F(n_2^2) = \left[\frac{(n_{\text{eff}})_0^2 \, \psi_1^2 - (n_{\text{eff}})_1^2 \, \psi_0^2}{(\psi_1^2 - \psi_0^2)} \right]$$

Equation 6.30 cannot be solved explicitly for n_2^2 but a solution is readily obtained by iteration. We make an initial guess for n_2^2, say $(n_2^2)_1$. This guess is substituted into the right side of Eq. 6.30 yielding a new estimate, $(n_2^2)_2 = F[(n_2^2)_1]$. The process is repeated and provided that

$$\left| \frac{\partial F(n_2^2)}{\partial \, n_2^2} \right| < 1$$

the iteration converges to the required solution. Proof of the convergence is given in Appendix 2.

PROBLEMS

6.1 Show that the transmissivity for the prism coupler given in Section 6.2,

$$t_d = \frac{k_{2x}}{k_{4x}} |T_{41}|^2 \, |T_{12}|^2 \, e^{-2\alpha_{1x}g}$$

can be simplified to yield the expression given in Eq. 6.12.

6.2 For the launching prism shown in Figure 6.20, what is the relationship between the angle θ_0, θ_i, and the bevel angle α for a prism of refractive index n_4?

6.3 A prism of refractive index $n_4 = 2.25$ is brought into proximity with a waveguide having film index $n_2 = 1.50$. The air-gap spacing between film and prism, g, is equal to 0.3164 μm. Make a plot of transmissivity t_d versus incident angle θ_i for a He–Ne laser beam with free space wavelength $\lambda_0 = 0.6328$ μm. Interpret your results.

6.4 The fundamental TE mode is excited by He–Ne laser on an asymmetric waveguide having substrate index $n_3 = 1.53$, film index $n_2 = 1.92$, and film thickness $d = 0.53$ μm. What is the effective waveguide thickness d_{eff}?

6.5 A prism having refractive index $n_4 = 2.58$ is used to couple a TE-polarized He–Ne laser beam into the waveguide described in problem 6.4.
   (a) What is the required angle of incidence θ_i to couple to most strongly to the fundamental TE waveguide mode?
   (b) If the gap spacing g equals 0.25 μm, what is the value of the transmissivity t_d through the prism coupler?

6.6 A thin film waveguide is characterized by the following parameters:

$$n_1 = 1 \qquad n_2 = 1.5 \qquad n_3 = 1.462 \qquad d = 0.9 \ \mu\text{m}$$

The fundamental TE mode of the waveguide is excited by a He–Ne laser, $\lambda_0 = 0.6328 \ \mu\text{m}$, through a prism having $n_4 = 2.25$.

(a) The effective refractive index n_{eff} for the above waveguide mode is calculated to the equal to 1.481. What is the corresponding synchronous angle of incidence θ_i?

(b) What is the gap spacing g required for a $1/e$ buildup distance z_n equal to 1 cm?

6.7 (a) For the waveguide and prism coupler described in problem 6.6 compute the bounce angle θ_p for the fundamental TE mode.

(b) Verify that the bounce above angle is correct by computing the quantity

$$- \phi_u(\theta) - \phi_t(\theta) + 2k_{2x}(\theta)d$$

in the weak coupling limit and showing that it is equal to zero for $\theta = \theta_p$.

(c) Make a plot of the deviation in phase, δ, versus θ as θ is varied about θ_p. Let the range of δ in your plot run from $-\pi$ to $+\pi$. What is the corresponding range in θ_i?

6.8 For the waveguide and prism coupler of problem 6.6, what should the prism length L be to maximize the amount of power coupled to the waveguide if the gap g is 0.3 μm?

6.9 (a) Which of the following equations can be solved by iteration? The solution x_0 for each is given.

1. $x = \cos x$ $x_0 = 0.739085$
2. $x = 3 \cos x$ $x_0 = 1.17012$
3. $x = 2 \tan^{-1} x$ $x_0 = 2.3312$
4. $x = \dfrac{\tan x}{2}$ $x_0 = 1.16555$
5. $x = e^{-x}$ $x_0 = 0.567143$

(b) Verify your answer by iteration on a calculator and indicate how many iterations are needed in each case for accuracy to three decimal places.

6.10 Using the film characterization technique described in Section 6.4, it is found that the first two modes of a particular dielectric slab waveguide are coupled out at the angles $\theta_i = 38.19$ and $34.26°$, respectively. The angle θ_i is defined as in Section 6.4. Write a program to iterate Eq. 6.30 thereby obtaining the film refractive index n_2 and thickness d. The other relevant waveguide parameters are as follows:

$$n_3 = 1.275$$
$$n_4 = 2.35$$
$$\lambda_0 = 1 \ \mu m$$

REFERENCES

1. Kapany, N. S., and Burke, J. J. *Optical Waveguides*. New York: Academic, 1972.

2. Tien, P. K. "Light waves in thin films and integrated optics." *Applied Optics* 10 (1971): 2395–2412.

3. Tien, P. K., Ulrich, R., and Martin, R. J. "Modes of propagating light waves in thin deposited semiconductor films." *Applied Physics Letters* 14 (1969): 291–294.

4. Ulrich, R., and Torge, R. T. "Measurement of thin film parameters with a prism coupler." *Applied Optics* 12 (1973): 2901–2908.

CHAPTER 7

Waveguide Fabrication

7.1 INTRODUCTION

The purpose of this chapter is to introduce the reader to a number of general methods used in fabricating two- and three-dimensional optical waveguides. The appropriate fabrication process for any given waveguide is a function of both its material composition and physical geometry. Four general classes of 2-D waveguides can be identified: the step-index and graded-index dielectric and the step-index and graded-index semiconductor waveguides. Although some overlap exists, each of these classes has its own set of processing techniques, as discussed in Sections 7.2–7.4. Fabrication of planar 3-D guides can be considered an extension of or modification of the processing techniques used for 2-D guides and is discussed in Section 7.5. Finally, fabrication methods for optical fiber waveguides are outlined in Section 7.6.

7.2 FABRICATION OF STEP-INDEX DIELECTRIC WAVEGUIDES

Step-index dielectric guides generally consist of a thin amorphous or polycrystalline guiding layer of high permittivity deposited on a relatively thick glass or crystalline substrate. The amorphous guiding layers can be made from glasses, polymer films, or metal oxides. When such a film is used in conjunction with a glass substrate, the waveguide thus formed generally acts simply as a passive means of transferring a guided signal between two points on the substrate. Alternatively, either the film or substrate or both can be made from crystalline material. Depending upon the application, the appropriate crystalline material can be chosen in such a fashion that its refractive index is sensitive to and therefore controllable by external application of either mechanical stress, electric or magnetic field, or a secondary optical signal. A number of such applications are discussed in ensuing chapters.

7.2.1 Sputtered-Film Waveguides

The process of sputtering may be defined as the ejection of atoms of molecules from a solid source target, in a vacuum, by the bombardment with atoms or ions having kinetic energies in the 10 eV to 2 keV range. The ejected atoms or molecules are allowed to impinge on a nearby substrate, producing a thin-film layer which is slowly built up with time as the individual particles are collected on the surface. Because the sputtering process can be mechanical rather than chemical in nature, it is well suited to the growth of a wide range of elements and compounds, independent of the constituent atom melting points and vapor pressures. The sputtering process generally produces very pure, uniform, durable, and optically low-loss amorphous or polycrystalline films.

In the glow-discharge sputtering method, shown schematically in Figure 7.1, the target material is placed at a negative potential (cathode) relative to the rest of the chamber which is assumed to be at ground potential. The substrate, on which the film is to be deposited, may be located above or to the side of the target, as shown. The entire assembly is placed in a vacuum ($\sim 10^{-6}$ Torr) and then the pressure in the enclosure is raised to the 10- to 20-m Torr range by introduction through an inlet valve of a flowing gas such as argon. In this pressure range, it is possible to ionize the gas by the application of several kilovolts to the target. The positively charged gas ions are accelerated toward the negatively charged target, transferring their momentum to it and thereby ejecting material that adheres to the substrate. The flux of positively charged ions from the discharge bombarding the target therefore causes sputtering of the target material. For insulating targets, the continuous flux of positively charged ions would rapidly increase its potential, thereby stopping the sputtering process. For these materials, an rf rather than dc voltage is applied to the target.

The sputtering deposition rate is a function of a number of parameters including target composition, ion kinetic energy, gas pressure and gas type or mixture, substrate-to-target separation, and substrate temperature. Sputtering rates are relatively low, being on the order of tens to a few hundreds of angstroms (Å) per minute for glass or metal oxide materials. However, if the sputtering conditions are held fixed, the rate of film growth is a linear function of time and can be accurately controlled. For nonelemental targets, the composition of the sputtered film is generally different from that of the target. This is a result of two factors. First, different atoms comprised by the target will have a different probability of being dislodged by an incident ion. As a result, some types of atoms in the target are more rapidly sputtered than others, yielding initially a gradual change in the target surface composition. This process continues until an equilibrium situation is reached between the ratio of the different types of atoms in the target and their respective sputtering rates.

Additionally, different ejected species may have different probabilities of adhering to the substrate, that is, different sticking coefficients. Further, the sticking coefficient can be dependent on substrate temperature, which is generally elevated by as much as

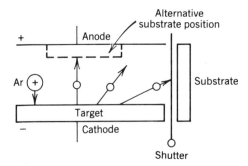

Figure 7.1 Geometry for glow-discharge sputtering.

200°C due to bombardment by electrons from the ionized gas. Because the substrate temperature may be a function of its composition, different substrates may yield films having different compositions, even when sputtered from the same target. For glass and metal oxide films, sputtering in an inert atmosphere often produces oxygen-deficient films. Proper stoichiometry can often be improved by the introduction of O_2 into the chamber.

To reduce the substrate temperature during sputtering, two modifications to the above process may be employed. In magnetron sputtering, a large dc magnet is incorporated, having magnetic field lines that are oriented in such a fashion as to confine electrons to the vicinity of the cathode. This has two effects: first, kinetic heating associated with electrons colliding with the substrate is reduced. Second, a greater proportion of gas molecules are ionized by the electrons located near the target rather than near the substrate. This results in a lower percentage of ionized molecules escaping from between the electrodes prior to impact with the target. Thus, shorter sputtering times are required for the same film yield. Alternatively, the glow-discharge plasma may be replaced with an ion gun source as shown in Figure 7.2. The gun produces a collimated beam of positively charged ions that is aimed at the substrate, producing sputtering. To prevent charge buildup on the substrate, the beam is first passed through a negatively charged grid that produces electrons that neutralize the beam. The beam is scanned across the substrate by the use of electrostatic deflecting plates as shown. Because of the greater isolation of the substrate from the ion generation processes, ion sputtering offers a number of advantages over conventional sputtering. These include control over substrate temperature, gas pressure, angle of deposition, and the type of particle bombardment of the growing film, as well as independent control over the ion current and energy.

A third variation on the glow discharge sputtering method is the reactive sputter technique. This process is useful for depositing metal oxide films such as TaO_5 and Nb_2O_5. For reactive sputtering, a metal target such as Ta or Nb is sputtered away in an oxygen-rich atmosphere and reacts chemically both at the target and substrate with the oxygen yielding a metal oxide film at the surface.

Sputter deposition, whether by glow discharge or ion beam, has been used to fabricate a wide number of different types of step-index waveguide geometries. These include both amorphous and polycrystalline films deposited on an amorphous substrate as well as amorphous films on a crystalline substrate. A representative sample of material combinations that have been fabricated are shown in Table 7.1 along with experimen-

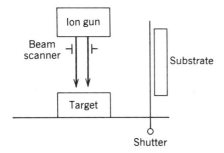

Figure 7.2 Geometry for ion-beam sputtering.

tally measured values of film and substrate refractive index and guided-wave attenuation.

7.2.2 Polymer-Film Waveguides

Several different techniques have been developed to deposit polymer films on amorphous substrates. In the solution deposit technique, the substrate is covered with a liquid such as photoresist, polyurethane, or mixtures of two polymer solutions such as polymethyl methacrylate (PMMA) and styrene acrylonitrile copolymer (SAN). The liquid layer thickness is controlled by either spinning the substrate at a rate of several thousand rpm in an axis perpendicular to its surface, vertically dipping the substrate into the liquid and removing it at a controlled rate, or covering the substrate with liquid and subsequently turning it upright to allow excess liquid to run off. Depending upon the polymer used, the films are subsequently air dried and baked at temperatures ranging from 60 to 100°C for times varying between 5 min to 70 h. The solution-deposit technique offers several potential advantages including simplicity of fabrication and required equipment, low cost, and low optical attenuation ($\sim$0.1–0.3 dB/cm). However, it is difficult to accurately control film uniformity and thickness.

A second method for the deposition of polymer films involves a plasma polymerization process. In plasma polymerization, an electrical discharge is created in a vapor containing low-weight organic molecules called monomers. The discharge causes an ionization and fragmentation of the monomers and a subsequent rebonding of the fragments into a much larger two- or three-dimensional structure by a process called cross-linking. When a substrate such as glass is introduced into the discharge, the polymer is deposited on its surface. Because the resulting polymer film has many chemical bonds, it is inert to most organic solvents, mild acids, and bases and exhibits good temperature stability.

The vacuum and discharge arrangement for plasma polymerization is very similar to that used in glow-discharge sputtering, except that no source target is used. Instead the monomer vapor is mixed with the argon gas flow stream, which is admitted to the vacuum chamber. The gas is ionized using an rf source applied between anode and cathode plates. The growth rate of the films is linear with time with typical rates on the order of 1000 Å/min. A number of different monomers have been used to fabricate polymer films using this technique. These include hexamethyldisiloxane (HMDS), vinyltrimethyl silazene (VTMS), cyclohexane, acetone, hexene-1, isopropyl alcohol, isopropyl silane, and perfluorocyclohexene. Extensive evaluation of the polymer film optical waveguide properties has been reported only for those made using HMDS and VTMS monomers. It was found that these guides exhibited extremely low losses of less than 0.04 dB/cm at 0.63 μm. By using a mixture of both HMDS and VTMS, the refractive index of the resulting films could be varied between 1.49 and 1.53. Further, it was found that the refractive index of these films could be modified after fabrication by heat treatment in an oxygen atmosphere. This feature is interesting in that it allows for the precise tuning of the refractive index required for optical components such as filters, resonators, and couplers. Table 7.2 lists the pertinent optical properties of a number of the polymer film waveguides discussed in this section.

Table 7.1 Composition and Optical Properties of a Number of Step-Index Dielectric Waveguides

Film Material	Film n	Substrate Material	Substrate n	Attenuation (dB/cm)
Barium silicate glass (a)	1.48–1.62 @ 0.63 μm	Fused quartz (a) Microscope slide (a)	1.512 @ 0.63 μm —	1.2
Ta_2O_5 (a)	2.2 @ 0.63 μm	Corning 7059 glass (a)	1.5285 @ 0.63 μm	0.9
SiO_2–Ta_2O_5 mixture (a)	1.46–2.08 @ 0.63 μm	Corning Vycor glass (a)	1.457 @ 0.63 μm	0.8
Nb_2O_5 (a)	2.1–2.3 @ 0.63 μm	7059 glass (a)	1.5285 @ 0.63 μm	1.0–2.0
Ta_2O_5 (a)	2.2 @ 0.63 μm	$LiTaO_3$ (x) with SiO_2 buffer layer	2.17, 2.18 (t)	1.0
ZnO (p)	—	Fused quartz (a)	1.512 @ 0.63 μm	—
GeO_2 (a)	1.6059 @ 0.55 μm	Microscope slide (a)	1.5158 @ 0.55 μm	0.7
Corning 7059 glass (a)	1.53–1.61 @ 0.63 μm	7059 glass (a)	1.5285 @ 0.63 μm	1.0

Notes. (a) Amorphous material; (p) polycrystalline material; (x) crystalline material; (t) retractive index depends on guided-mode polarization.

175

Table 7.2 Optical Properties of a Number of Polymer Film
Step-Index Waveguides

Film type	Refractive index	Attenuation (dB/cm)
Polystyrene	1.5860, TE modes	—
	1.5888, TM modes	
PMMA–SAN mixture	1.490–1.565	0.2
Polyurethane		
Type 9653-1	1.555	0.8
Type LX500	1.573	4
Epoxy	1.581	0.3
Photoresist (KPR)	1.615	7.0
VTMS–HMDS mixture	1.488–1.528	0.04–0.3

7.3 FABRICATION OF GRADED-INDEX DIELECTRIC WAVEGUIDES

Graded-index dielectric waveguides are generally produced by the modification of the refractive index of a crystal or amorphous insulating substrate near its surface. This can be accomplished through the introduction of a foreign species into the surface region by either thermal in-diffusion or simple kinetic impact, by exchange of some of the constituent atoms in the substrate material near its surface with ionic species located above the surface, or by removal of certain atomic species from the surface region of the substrate by thermal diffusion. All of the above techniques yield a refractive-index profile that varies in a continuous rather than stepwise fashion with depth into the substrate.

7.3.1 Ion-Migration Waveguides

For a number of optical communication applications, it is desirable to design wave-guiding structures on a substrate which interfaces with incoming and outgoing multimode optical fibers. For efficient transfer of energy from the fibers to the substrate and back again, fiber and substrate permittivities should be similar and the width of the guiding region in the substrate should be comparable to the diameter of the guiding core of the fiber, as shown in Figure 7.3. Since, as was demonstrated in Chapter 4, multimode operation implies a wide guiding region and because most multimode fibers are made from glass, the above constraints for efficient coupling translate into the need for glass substrates with large guiding regions. Ion migration provides a method for fabricating such fiber-compatible structures.

Many of the glasses used in OIC applications are composites consisting of mixtures of SiO_2 with metal oxides such as Li_2O, Na_2O, Al_2O_3, and K_2O. The refractive index

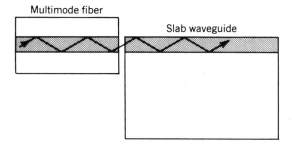

Figure 7.3 Efficient coupling between an optical fiber and a planar waveguide using guiding regions of comparable permittivity and physical dimensions.

of glass is related both to its density and to the electronic polarizability of the constituent ions. By diffusing ions such as Ag^+, Tl^+, and K^+ into the glass, which replace lighter ions such as Na^+ and Li^+, the diffused region can be made to have a higher refractive index than that of undiffused glass.

As a simple example of the ion-migration process consider a soda-lime glass, which contains approximately 13% of Na_2O by weight, immersed in a molten bath of the salt $AgNO_3$ as shown in Figure 7.4. The upper surface of the glass substrate has an aluminum contact evaporated on it and is connected to ground potential. The molten salt is held at a positive potential by means of an immersed electrode. Typical melt temperatures for silver salt range between 200 and 350°C with applied electric field strengths in the $0–5 \times 10^5$-V/m range. Alternatively, the molten metal salt may be replaced by a thin evaporated Ag layer which is over-coated with a thin aluminum film that acts as the anode. The substrate is subsequently placed in an oven and heated as described above.

Consider first the situation with no applied voltage. Because of the concentration gradient, Ag^+ ions diffuse into the surface of the substrate while Na^+ ions migrate out. The presence of diffused Ag^+ near the substrate surface produces an increase in refractive index in this region. The increase in refractive index near the surface using this method is about 0.1. Propagation losses are on the order of a few tenths of a decible per centimeter. The diffusion time required to fabricate multimode guides using

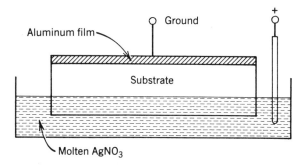

Figure 7.4 Simple setup for creating a waveguide by ion migration.

this method is very large; for example, a guide capable of supporting 13 modes requires diffusion for 6 h at 250°C. With the application of an electric field, diffusion times can be drastically reduced. Depending on the applied electric field and substrate temperature, a reduction in diffusion times by a factor between 5 and 3000 is possible. Figure 7.5 shows the dependence of field-aided diffusion on applied voltage for an evaporated silver film on soda-lime glass. It should be noted that the functional form for the change in refractive index with depth is generally not describable in terms of the usual Gaussian or complementary error functions due to (1) the presence of the field-aided diffusion term, (2) the different diffusion coefficients for the various species, and (3) a diffusion coefficient that varies with depth.

A two-step migration process has also been demonstrated to create buried-channel guiding structures. The first step of the process is similar to that described above. A molten mixture of thallium, sodium, and potassium salts is placed in contact with a borosilicate glass that contains potassium oxide and sodium oxide. In the first step, high-polarizability thallium (Tl^+) ions diffuse into the glass surface creating a region of higher permittivity. In the second step, a melt containing only the sodium and potassium salts is used. This results in a diffusion of less polarizable Na^+ and K^+ ions into the surface and a further diffusion of Tl^+ deeper into the substrate creating a region of high permittivity embedded in lower permittivity regions. Measured values for maximum change in refractive index using this method are on the order of 0.003.

7.3.2 Proton-Exchange Waveguides

Because of the acousto-optic, electro-optic, and piezoelectric properties of lithium niobate and lithium tantalate, techniques that allow fabrication of waveguides using these single-crystal materials as substrates are of great interest. One such method is proton exchange.

Proton exchange is a technique that is similar to ion migration in which metal ions

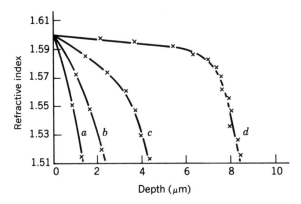

Figure 7.5 Field dependence of the index profiles at $T = 250°C$, $t = 30$ min, for various voltage values. (a) 10 V; (b) 30 V; (c) 50 V; (d) 100 V. (After Chartier et al. (Ref. 10). Reproduced by permission of the Institution of Electrical Engineers.)

in a crystalline substrate are partially replaced by hydrogen ions. This approach is found to produce large changes in refractive index, having a nearly steplike profile in the surface regions of $LiNbO_3$ and $LiTaO_3$ substrates. The procedure entails placing the substrate in a solution of benzoic acid which is heated to around 200°C. In equilibrium, the acid dissociates as follows:

$$C_6H_5COOH \longleftrightarrow C_6H_5COO^- + H^+$$

The amount of dissociation and thus the number of protons depends only on temperature. A surface guiding region is formed because protons from the acid exchange with Li ions under the equilibrium reactions:

$$LiNbO_3 + xH^+ \longleftrightarrow Li_{1-x}H_xNbO_3 + xLi^+$$

and $\hspace{10cm}$ (7.1)

$$LiTaO_3 + xH^+ \longleftrightarrow Li_{1-x}H_xTaO_3 + xLi^+$$

Because of the crystalline nature of the substrate, the optical properties of materials such as lithium niobate and tantalate are dependent on their orientation. For these materials, proton exchange has significant influence only upon the refractive index seen by waves having electric field polarization components along the crystallographic Z axis. For such a polarization, the increase in refractive index is about 0.12 at 0.63 μm. Further, the process is exceedingly rapid. At 249°C, a single-mode waveguide is fabricated in less than 5 min and a surface guiding region 10 μm thick is formed after 24 h. Propagation losses are as low as 0.5 dB/cm.

When pure benzoic acid is used, as described above, the process has been reported to replace as much as 72% of the Li ions in the crystal surface. These high proton concentrations are presently believed to cause damage to the crystal surface. To reduce the amount of proton exchange, lithium salts such as $LiNO_3$, Li_2CO_3 are added to the melt. These salts dissociate in the melt as follows:

$$LiNO_3 \longleftrightarrow Li^+ + NO_3^-$$

$$Li_2CO_3 \longleftrightarrow 2Li^+ + CO_3^-$$

The dissociation process therefore introduces additional Li^+ ions thereby pushing the equilibrium exchange process described by Eq. 7.1 to the left. Further, the NO_3^- and CO_3^- ions combine with benzoic acid via the reactions

$$NO_3^- + C_6H_5COOH \longrightarrow HNO_3 + C_6H_5COO^-$$

and

$$CO_3^- + 2(C_6H_5COOH) \longrightarrow H_2CO_3 + 2 C_6H_5COO^-$$

with the nitric and carbonic acid boiling away. This reduces the amount of protons available for exchange.

7.3.3 Waveguides Formed by Metal In-Diffusion

Metal in-diffusion is another fabrication technique that is employed to form waveguides on single crystal dielectric substrates. In this method a thin metallic film of thickness ranging from 150–1500 Å is sputtered or evaporated onto the surface of the substrate crystal. The sample is then placed in a quartz oven and heated in inert or a combination of inert and oxygen atmosphere at temperatures around 1000°C for a period of several hours, thereby diffusing the metal into the surface region of the substrate. Such a diffusion produces a Gaussian distribution of metal atoms within the lattice. The presence of these atoms creates an increase in refractive index that is roughly proportional to their concentration. Thus, due to the Gaussian distribution of atoms with distance x into the substrate, the refractive index is given by

$$n(x) = n_3 + \Delta n e^{-x^2/d^2} \tag{7.2}$$

where n_3 is the substrate index and the constant d is related to the diffusion time t and diffusion coefficient D by

$$d = 2\sqrt{Dt}$$

This variation in refractive index is shown in Figure 7.6.

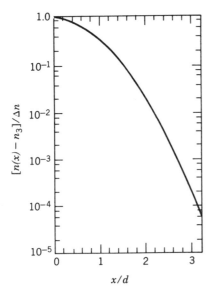

Figure 7.6 Variation in refractive index with distance x for metal in-diffusion. (Edward S. Yang, *Fundamentals of Semiconductor Devices.* Copyright © 1978, McGraw–Hill, New York. Reproduced by permission.)

The above method has been used to fabricate waveguides on both $LiTaO_3$ and $LiNbO_3$ substrates. For lithium tantalate, the most widely used metal is niobium, which is diffused into the substrate in an argon atmosphere. The niobium atoms go into the lattice substitutionally, replacing some of the substrate Ta atoms and forming $LiNb_xTa_{1-x}O_3$. Maximum changes in refractive index using this method are on the order of 1–3% at He–Ne wavelengths. This method can be used to fabricate single-mode waveguides with guiding regions 1–2 μm in thickness and low propagation losses of less than 1 dB/cm.

Similar techniques have been reported for forming waveguides on $LiNbO_3$ substrates using titanium atoms as the diffusant. For these guides the substrate is exposed to water vapor and to O_2 during a portion of the oven cool-down cycle to prevent out-diffusion of Li_2O.

7.3.4 Ion-Implanted Waveguides

When a stream of energetic ions impinge upon a solid substrate, these particles lose their kinetic energy through a series of collisions with substrate atoms and electrons, eventually coming to rest within the material. This process is known as ion implantation. Ion implantation can be used to fabricate optical waveguides both in semiconducting and dielectric substrates by creating narrow regions of high permittivity at or near the substrate surface.

Figure 7.7 shows the basic ion-implantation apparatus. The ion source supports a gas discharge that may consist primarily of the desired implant species, for example argon, or may use an inert gas to create the discharge. In the latter case, the desired species is obtained by its introduction into the discharge in the form of vapor or sputtered metal and its subsequent ionization by electrons within the gas discharge. The generated ionic species diffuse out of the source and are then accelerated between high-voltage plates to energies typically in the 10- to 500-keV range. The various species in the ion beam are separated spatially according to their mass through the use of a magnet. Due to the interaction of the magnetic field with the moving ions, each species travels through the separator at a radius that depends on its mass. By adjusting the magnetic field strength properly, only the desired species is directed toward the target. The position of ion beam can be scanned across the target through the use of electrostatic deflection plates.

In dielectric substrates, the modification in permittivity is primarily due to (1) damage in the form of displaced atoms occurring as the result of ion collisions with atoms in the substrate and (2) ionization of substrate atoms leading to structural defects such as broken bonds. Ionization damage tends to be localized near the crystal surface where the incident particles still have high energy. Atomic collision damage occurs further into the substrate bulk where the particles have lost a sufficient amount of kinetic energy to cause atomic displacement. Because the atomic collision process occurs near the final resting point of the implanted ions, it is not surprising that the spatial distribution of this damage is often similar to that of the actual implanted ions. Figure 7.8 shows the variation in refractive index measured for He^+ implants into fused SiO_2 as a function of incident ion energy E and dose N_d. Dose is defined as the total number

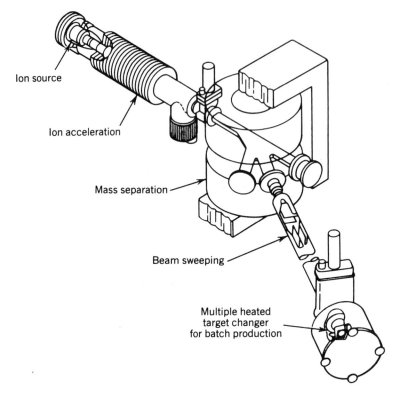

Ion source

Ion acceleration

Mass separation

Beam sweeping

Multiple heated
target changer
for batch production

Figure 7.7 Components of a typical ion implanter. (J. W. Mayer, L. Eriksson, J. A. Davies, *Ion Implantation in Semiconductors.* Copyright 1971, Academic Press. Reproduced by permission.)

of ions per unit area that are incident upon the substrate. For this material, damage results in an increase in refractive index. Note from Figure 7.8*b* that for high doses, the refractive index distribution approaches that of a step waveguide of depth *d* where from Figure 7.8*a, d* increases approximately linearly with ion energy. Further, for low ion doses, the electronic ionization contribution to modification of the refractive index is small. In this limit, the distribution for amorphous targets can often be approximated by a function of the form

$$C(x) = \frac{N_{\mathrm{d}}}{\Delta R_{\mathrm{p}}(2\pi)^{1/2}} \exp\left[\frac{-(x - R_{\mathrm{p}})^2}{2\Delta R_{\mathrm{p}}^2}\right] \qquad (7.3)$$

where $C(x)$ is the implanted-ion concentration per unit volume as a function of distance x into the substrate, N_{d} is the ion beam dose, R_{p} is the average penetration depth, known as the projected range, and ΔR_{p}, called the range straggling, measures the amount spread of the ion distribution about its average value. This type of distribution is shown diagrammatically in Figure 7.9.

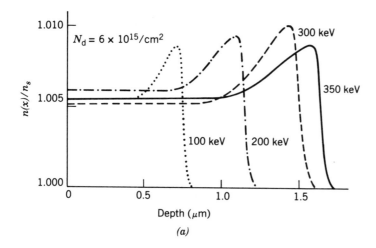

(a)

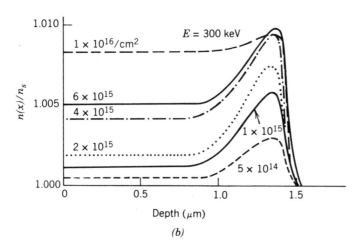

(b)

Figure 7.8 Refractive-index dependence of fused silica on He$^+$ implant dose and energy. (After Heibei *et al.* (Ref. 29). Reproduced by permission of *Physica Status Solidi.*)

As might be anticipated by the previous discussion of the collision process, the range and straggling coefficients are a function of the substrate material, e.g., substrate density and mass of the constituent atoms, ion type, and ion energy. For fixed ion and target type, therefore, ion beam energy and current, or flux, can be used to control the depth and width of the implanted region.

Ion-implanted waveguides in fused silica have been fabricated using a wide range of ions, including He, Li, C, P, Xe, and Te, although the most complete studies have been performed using Li ions. These light ions produce less surface damage than other species, thereby reducing optical scattering losses, and also have a relatively long range which is convenient in forming guiding layers of the desired thickness. For these ions,

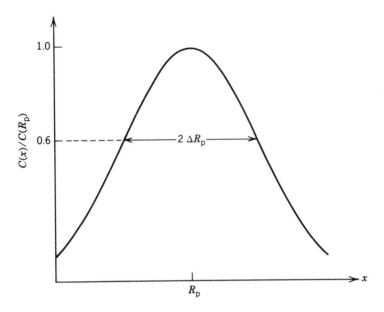

Figure 7.9 Typical Gaussian implant distribution showing range and straggling.

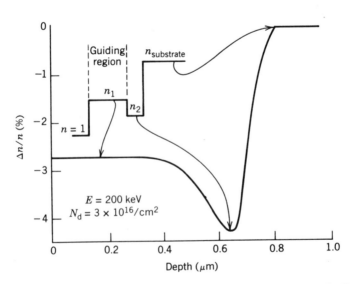

Figure 7.10 Index profile for a waveguide formed by He$^+$ implant in LiTaO$_3$. E = 200 keV, dose = 3 × 10^{16} cm^{-2}. (After Wenzlik *et al.* (Ref. 79). Reproduced by permission of *Physica Status Solidi*.)

the refractive index varies in an approximately linear fashion as

$$n = 1.458 + 2.1 \times 10^{-21} C \qquad (7.4)$$

where C is the concentration of implanted ions per cubic centimeter. Both surface and buried waveguides have been made using this technique and exhibit a propagation attenuation estimated at 5 dB/cm at a wavelength of 0.63 μm. Recently, extremely low-loss waveguides (0.1 dB/cm) have been formed by implantation of nitrogen ions into fused silica. For these waveguides, the resulting increase in refractive index appears to be chemical in nature with significant quantities of nitrogen dissolved in amorphous SiO_2 to form the higher refractive index SiO_xN_y.

Ion implantation using ions such as N, O, B, Ne, H, and He has also been used to create waveguiding in $LiNbO_3$ and $LiTaO_3$. For these materials, damage creates regions of *lower* permittivity. This yields the type of waveguide shown by the permittivity profile of Figure 7.10. The step-index approximation to this profile is shown in the inset.

7.4 STEP-INDEX SEMICONDUCTING WAVEGUIDES

It is often desirable to integrate passive optical waveguide components with either optoelectronic devices such as solid state laser sources, photodetectors, or optical modulators, or to include the capability for electronic amplification or signal processing. This can be accomplished by fabricating waveguides out of semiconducting materials. Again, depending upon the fabrication method and starting materials, either step- or graded-index profile waveguides can be made. Because of the progress made in growth techniques, step- rather than graded-index semiconducting waveguides are used almost exclusively.

The structure of step-index semiconductor waveguides can vary greatly depending on the application. For the simplest geometry, a high-permittivity single-crystal semiconducting guiding layer is grown epitaxially on a lower permittivity substrate. More complicated waveguiding geometries are also employed in semiconductor optoelectronic components such as diode lasers wherein a series of epitaxial semiconducting layers of different doping and alloy composition are built up on a substrate. By controlling these parameters, not only can the permittivity profile be tailored but also important optoelectronic properties such as band gap, carrier confinement, carrier transport, and optical absorption.

The two primary semiconducting materials used in integrated optics are $Ga_{1-x}Al_xAs$ and $Ga_{1-x}In_xAs_{1-y}P_y$, where x and y represent the mole fractions of the various indicated constitutent components. Variation of the alloy composition is the primary mechanism for control of refractive index. For $Ga_{1-x}Al_xAs$ waveguides, use is made of the fact that the refractive index decreases with increasing mole fraction of Al, varying from $n \cong 3.57$ for $x = 0$ (GaAs) to $n \cong 3.35$ for $x = 0.35$ at $\lambda_0 = 0.9$ μm (larger values of x have deleterious effects and are not generally used). Figure 7.11 shows an example of such a waveguide along with the refractive index profile for $\lambda_0 = 0.85$ μm. Note

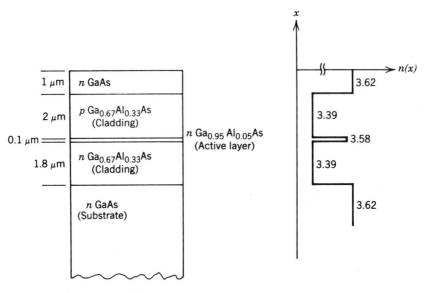

Figure 7.11 GaAlAs waveguide.

that provided the evanescent fields do not extend appreciably from the core into the cladding region, the waveguide can be modeled to good approximation as a slab structure. For $Ga_{1-x}In_xAs_{1-y}P_y$, although there are two independent variables, in order for lattice matching with the substrate which is usually InP, it is required that $y = 2.197x$. Figure 7.12 shows the dependence of refractive index on y for various values of λ_0.

One additional method exists for forming guiding layers in semiconducting waveguides, which makes use of the lowering of the refractive index by the addition of free carriers to the material. An example of this type of structure is shown in Figure 7.13 where a GaAs waveguide is formed by the epitaxial growth of a lightly doped n^- film on top of a more heavily doped n^+ GaAs substrate.

7.4.1 Liquid-Phase Epitaxial Film Waveguides

Liquid-phase epitaxy (LPE) is the technique by which an epitaxial crystalline layer is grown on a single-crystal substrate by solidification of a molten solution that is saturated at the growth interface. The LPE method is used extensively in the fabrication of thin, single-crystal semiconducting layers of the ternary alloys of GaAlAs on GaAs substrates and quaternary alloys of GaInAsP on InP substrates. These materials are particularly important for lasers, detectors, and OIC applications in the areas of high-speed optical signal processing and optical fiber communications. In addition to the fabrication of semiconductor films, LPE provides a means for growing thin films of a number of single-crystal insulating materials such as $LiTaO_3$ and $LiNbO_3$.

The primary factors controlling the properties of LPE grown films are substrate orientation, melt and substrate temperatures, melt composition, cooling rate, and growth

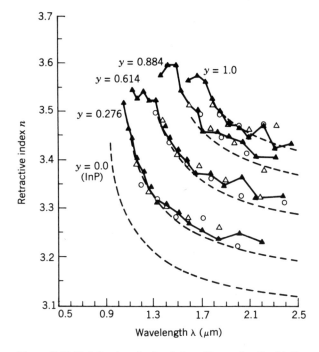

Figure 7.12 Variation in refractive index with wavelength of lattice-matched $Ga_{1-x}In_xAl_{1-y}P_y$ for various values of y. (After Chandra *et al.* (Ref. 6). Reproduced by permission of the Institution of Electrical Engineers.)

time. By accurate control of these parameters, repeatable films of precise composition and thickness can be made. An important limitation of the LPE techniques is that to obtain good quality films the mismatch in lattice constants between film layers and the substrate should typically be less than 1–2%. In addition, LPE layers often do not exhibit as good a surface quality as layers formed by vapor-phase epitaxy (VPE). The latter technique is discussed in Section 7.4.2.

Growth of films by LPE is based upon the fact that the solubility of one or more atomic species dissolved in a solution decreases as the solution temperature is lowered. When such a solution is placed in contact with a substrate and is cooled, condensation of the dissolved species (known as the solute) onto the substrate occurs, resulting in

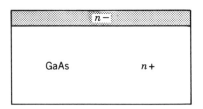

Figure 7.13 GaAs waveguide formed by epitaxial growth of lightly doped n^- film on top of a doped n^+ substrate.

film growth. Although a number of different techniques exist for film growth, three principle methods are used: tipping, dipping, and sliding.

In the tipping technique, illustrated in Figure 7.14, the substrate is held tightly against the bottom of the upper end of a graphite boat and the solution containing the material to be deposited is placed at the lower end. The boat is placed in a quartz furnace tube with H_2 flowing through the tube. The boat is then heated to the desired growth temperature. The oven temperature is subsequently reduced at a suitable cooling rate and the furnace is then tipped, permitting the solution to flow over the substrate. At the time of tipping, the temperature of the solution is such that is saturated or nearly saturated with the growth material.

Upon further cooling, the material percipitates out of solution forming a layer on the substrate surface. After a cooling time Δt the furnace is tipped back to its original position.

The dipping technique is similar to the tipping method except that the substrate is lowered into a crucible containing the molten solution to initiate growth and is raised to its original position to terminate contact. This method has been used not only to grow semiconducting films but also to produce a layer of $LiTaO_3$ on a $LiNbO_3$ substrate.

Perhaps the most widely used method is the sliding technique, which permits the growth in succession of a number of different alloy films. As shown in Figure 7.15, the multiple-bin graphite boat consists of a graphite block having a number of cylindrical bore holes that act as reservoirs for the various growth solutions. The graphite block is provided with a thermocouple for measurement of the growth solution temperature and a graphite cover for the bins. A very flat graphite slider is positioned below and acts as the bottom for the reservoirs. Recessed into the slider is a shallow depression which holds the substrate. The entire boat is located in a quartz tube through which highly purified H_2 flows.

Growth by the slide technique is accomplished by heating the melt material to the desired temperature. A particular furnace cooling rate is then initiated and the substrate is slid under the first bin using a quartz push rod. When the appropriate growth time has been reached, the substrate is slid underneath the next bin and the second layer grown. Growth times for these layers can be only seconds with thicker layers formed in minutes.

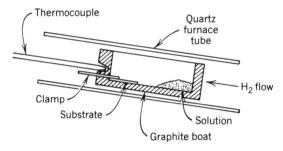

Figure 7.14 LPE by the tipping technique. (After Nelson (Ref. 50). Reproduced by permission of *RCA Review.*)

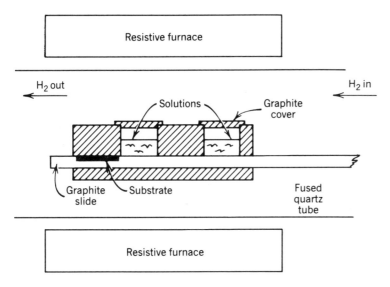

Figure 7.15 LPE by the slider method.

To better understand the mechanism for LPE film growth let us consider the growth of a thin layer of GaAs on a GaAs substrate. Figure 7.16 shows the solubility x_{AS} of the atom fraction of As in metallic Ga as a function of temperature. For example, at 800°C x_{AS} is approximately 2%. Thus, a 98% atomic fraction of Ga and 2% atomic fraction of As can be melted at temperatures slightly above 800°C. As the temperature is slowly reduced below this point, excess As must precipitate out of solution. It does so by combining in a stochiometric fashion with Ga, that is, one atom of As with one atom of Ga from the melt, forming a layer of GaAs on the surface. Note that the small amount of Ga used up in the melt is a negligible fraction of the total amount available so that its removal from the melt has a minimal effect on the ratio of Ga to As. Note further that GaAs has formed from a melt having a temperature which is far below that required to melt an ingot of GaAs (1238°C). The above technique can be extended to grow alloys such as $Ga_{1-x}Al_xAs$ where x represents the mole fraction of Al in the film. This is accomplished by initial choice of the atomic percent of Al and As in the melt. Similar techniques permit controlled growth of the alloy $Ga_{1-x}In_xAs_{1-y}P_y$.

7.4.2 Chemical Vapor Deposition Epitaxial Waveguides

Chemical vapor deposition (CVD), also called vapor-phase epitaxy (VPE), refers to the process of growing crystallographically oriented films on a substrate by the reaction of chemical vapors at high temperatures. The popularity of the CVD processes in the past 20 years can be attributed to its wide flexibility in terms of chemical compositions that may be grown, the flexibility in control of film composition, as well as its compatibility with large-scale commercial processing. The two most widely used CVD

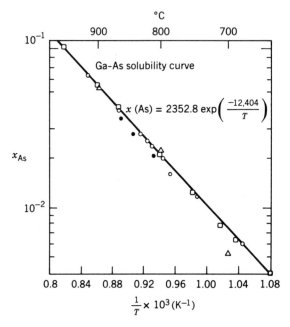

Figure 7.16 Ga–As solubility curve. (After Hsieh (Ref. 32). Reproduced by permission of the author and North-Holland Physics Publishing.)

epitaxial growth procedures are hydride/chloride CVD for GaInAsP compounds, and metal–organic CVD (MOCVD) for both GaInAsP and GaAlAs compounds.

Epitaxial film growth of GaInAsP alloys using the hydride/chloride technique is a two-step process. A typical CVD growth system is shown in Figure 7.17. In the first step, HCl gas is passed over hot (800–850°C) Ga or In metal, forming metal chloride vapor and hydrogen gas via the reactions

$$Ga(s) + HCl(v) \longrightarrow GaCl(v) + \tfrac{1}{2}H_2(v)$$

$$In(s) + HCl(v) \longrightarrow InCl(v) + \tfrac{1}{2}H_2(v)$$

The metal chlorides are then combined with hydrides of As (AsH$_3$) and P (PH$_3$) in a mixing chamber (825–900°C). The hydrides first dissociate or ''crack'' at high temperatures via the reactions

$$AsH_3 \longrightarrow \tfrac{1}{2}As_2 + \tfrac{3}{2}H_2$$

and

$$PH_3 \longrightarrow \tfrac{1}{2}P_2 + \tfrac{3}{2}H_2$$

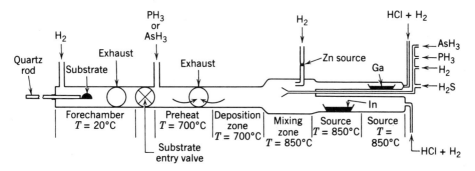

Figure 7.17 VPE furnace for growth of GaInAsP. (T. P. Pearsall, *GaInAsP Alloy Semiconductors*. Copyright © 1982, John Wiley & Sons, Ltd. Reprinted by permission.)

Free As and P then combined with GaCl and InCl vapors in a cooler deposition zone (650–725°C) to form a solid $Ga_xIn_{1-x}As_yP_{1-y}$ alloy. The released chlorine combines with H_2 yielding HCl as a by-product. Figure 7.18 shows a proposed model for the nucleation process. Typical growth rates range from 1 to 10 μm per hour. By adjusting oven temperature and gas flow rates, the alloy composition can be precisely controlled. N-type doping using sulfur can be accomplished by adding H_2S gas in the mixing zone as shown. Zn vapors produced by passing H_2 over hot Zn metal can be used for P-type doping.

An alternative method for the formation of both III–V and II–IV semiconductor films from the vapor phase is MOCVD. The process is described with reference to the growth of GaAlAs alloys, although the general procedure is similar for other compounds.

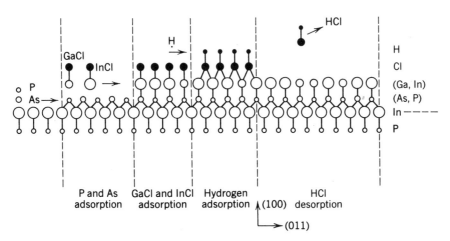

Figure 7.18 Proposed model for nucleation of GaInAsP. (T. P. Pearsall, *GaInAsP Alloy Semiconductors*. Copyright © 1982, John Wiley & Sons, Ltd. Reprinted by permission.)

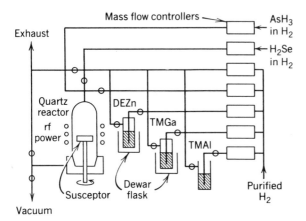

Figure 7.19 GaAlAs MOCVD reactor. (After Dupuis *et al.* (Ref. 15). Copyright 1979, IEE. Reproduced by permission.)

A schematic diagram of a GaAlAs MOCVD reactor is shown in Figure 7.19. The growth chamber consists of a quartz tube reactor containing a SiC susceptor that holds the substrates on which the films are to be deposited. The susceptor is heated inductively by an external RF source. The metal–organic vapors, trimethylgallium (TMGa), trimethylaluminum (TMAl), and arsene (AsH$_3$) are used as sources. These compounds are pyrolyzed in an H$_2$ atmosphere at 650–750°C to form thin films by the following reaction:

$$(1 - x)[\overbrace{(CH_3)_3Ga]}^{\text{TM Ga}} + x[\overbrace{(CH_3)_3Al]}^{\text{TM Al}} + AsH_3 \xrightarrow[\text{heat}]{H_2}$$

$$Ga_{(1-x)}Al_xAs + 3CH_4 \text{ (vapor)}$$

The composition of the film is thus controlled by the relative partial pressures of the source components, TM Ga and TM Al. When doping is desired, the gas diethyl zinc (DEZn) can be used for P-type Zn doping and the gas hydrogen selenide (H$_2$Se) for N-type selenium doping.

Because the metalorganics are liquids near room temperature with high vapor pressures, they are vaporized by bubbling H$_2$ through them which is used as the carrier gas. The relative flow rates for the constituent gases can be precisely controlled using mass flow meters. Various layer thicknesses and compositions can be controlled by a timed sequencing of appropriate gas flow paths. Typical film growth rates are on the order of 3000 Å/min.

7.4.3 Molecular-Beam Epitaxial Waveguides

Molecular-beam epitaxy (MBE) is the process by which epitaxial films are grown from beams of atoms or molecules by their reaction with a crystalline surface under ultrahigh vacuum conditions. The molecular beams are thermally generated in small ovens known as Knudsen-type effusion cells, each of which produces molecular fluxes that are controlled by orifices or shutters. The flux intensity is determined by the temperature of the material in the effusion cell and the distance from the cell. For a fixed temperature and distance from the cell, the flux intensity of the molecular components of the beam is fixed and depends only on the partial pressures of the constituent source atoms being vaporized. These pressures are a measure of how easily the individual types of atoms are vaporized from the heated source material and at a fixed temperature are only a property of the source composition.

By using a number of effusion cells, each containing a different molecular beam source material, and combining the respective beam fluxes at the substrate, a wide variety of film compositions may be created. These compositions are controlled by the choice of the particular source material used in each oven as well as by thermal control of the relative flux intensities. Oven beams can be rapidly turned on and off and because, depending on the material, growth rates on the order of 1–10 monolayers per second can be maintained, layer thickness as low as 10 Å can be grown. This feature has been demonstrated by the growth of "superlattice" structures consisting of alternating layers of GaAs and $Ga_{1-x}Al_xAs$ with thicknesses as low as 10 Å. The ability to grow such thin layers is particularly useful for fabrication of some of the more sophisticated diode lasers presently made on both GaAlAs and GaInAsP alloys. The precise control over layer thickness and composition is an advantage offered by MBE over LPE and CVD techniques. A primary disadvantage is the difficulty in scaling up the process to permit large-scale production.

The apparatus for MBE growth is shown schematically in Figure 7.20. The essential features include an ultra high vacuum bell jar, typically made from stainless steel, one or more effusion cells, molybdenum heating block, substrate, ion-sputtering gun, and

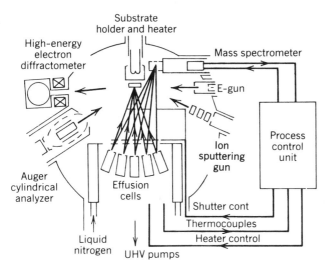

Figure 7.20 MBE configuration. (After Chang (Ref. 7). Reproduced by permission of the author and North-Holland Physics Publishing.)

a number of pieces of analysis equipment for providing information on the chemical composition of the substrate surface and grown film. The ovens themselves are generally made from boron nitride or high-purity graphite. They contain an inner crucible and outer tube that is wound with resistive heating wire. Each oven is thermally isolated from the remainder of the vacuum chamber environment by a liquid-nitrogen-chilled shroud, which also acts to condense out residual gas contaminants in the vacuum system, and is provided with a thermocouple and mechanical shutter.

In operation, the system is first evacuated to pressures on the order of 2×10^{-9} Torr and the chamber walls are subsequently baked out for a number of hours to outgas residual gas contaminants yielding final chamber pressures of about 1×10^{-10} Torr. Residual gas contamination is monitored with the mass spectrometer. Prior to film growth, the substrate is cleaned by sputtering, using the ion-gun source which removes a surface layer of the substrate along with any associated contamination. Substrate surface composition is monitored with an Auger electron spectrometer (AES). This device bombards the substrate surface with a beam of energetic electrons causing valence or conduction electrons to be ejected. The intensity and energy at which these electrons are ejected provides a distinct signature as to the types and quantities of the constituent atoms in the surface. Auger spectroscopy can also be used to monitor film composition during and/or after growth. In addition to information on film composition, the stoichiometry of the grown film can be evaluated by reflection high-energy electron diffraction (RHEED) measurements. Here energetic electrons impinge upon the film surface at glancing angles and the scattered radiation is imaged as a diffraction pattern on a fluorescent screen.

7.5 THREE-DIMENSIONAL WAVEGUIDES

7.5.1 Raised-Channel Waveguides Formed by Mask and Etch

One of the principal methods for the fabrication of strip, rib, and strip-loaded waveguides is the mask and etch process, shown in Figure 7.21. The central guiding rib or strip is produced by covering the surface of the 2-D guide with a protective mask having a width equal to that of the desired rib or strip and subsequently etching away material to either side. By depositing an additional film layer over the strip, embedded strip guides can also be fabricated. Masks may be made from either developed photoresist or metal. Mechanical or "dry" etching is generally accomplished by bombarding the substrate film surface with ions in either an inert or reactive gas atmosphere. These ions cause surface material to be removed either by mechanical sputtering or as a result of chemical reaction of the ions with the surface. "Wet" etching is accomplished by placing the exposed surfaces in contact with a solvent which removes material by chemical reaction. Several of the most widely used methods will be described in detail below.

Waveguides using photoresist masking are fabricated using the following procedure: the 2-D guide is spin-coated with a layer of liquid photoresist to a thickness of several thousand angstroms. The resist is subsequently hardened by baking at approximately

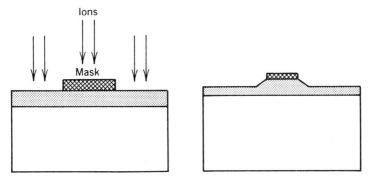

Figure 7.21 3-D waveguide formed by dry etching.

90°C for about 30 min. The photoresist pattern can be delineated by either exposure to UV light through a photomask or by exposure to a scanning electron beam which "writes" the desired pattern on the resist. After the exposure, the photoresist is developed, leaving the desired mask above the waveguide. Because of the higher spatial resolution of the electron beam, masks made using this technique generally have smoother, more sharply defined walls than those made by optical exposure. Consequently, after etching, the walls of the raised portion of the rib or strip will also be smoother. The nature of wall surface is of great importance because roughness introduces scattering from the guide, resulting in attenuation of the guided wave as it propagates.

Dry etching can be accomplished using either the glow-discharge or ion-gun sputtering techniques described in Section 7.2.1 with the 2-D guide now acting as the target. Ion-gun sputtering is generally preferable because (1) it provides more control over the direction of incident ions relative to the substrate surface thereby permitting more flexibility in determining the profile of the etched rib or ridge, (2) it is "cleaner" since etching can be performed in high vacuum and sputtering from the walls and fixtures of the vacuum chamber is eliminated, and (3) it allows better control of the substrate temperature since the substrate is not immersed in a plasma as is the case with glow-discharge etching.

The etching technique described above suffers from several problems that often result in unsatisfactory guiding walls. First, because the sputtering is a mechanical process, both the substrate as well as the mask are gradually removed. Both the height and the width of the mask decrease as the etching proceeds resulting in tapered rather than vertical walls. It should be pointed out that for some applications such tapering may be desirable, but often it is not. An additional drawback to the sputtering technique is redeposition of sputtered material on the previously etched raised portions of the waveguide walls. This can result in rough, nonvertical walls and consequently lossy guides.

An alternative dry-etch technique which addresses the etching problems described above is reactive sputter etching. Reactive sputter etching is an extension of the glow-discharge method in which chemically active gas is introduced between the rf discharge

13106 15KV 50U

Figure 7.22 Reactively etched glass waveguide. (After Izawa *et al.* (Ref. 34). Reproduced by permission of the authors and the American Institute of Physics.)

plates. The type of gas is chosen so that when ionized, one of the generated constituent ions is primarily chemically reactive with the substrate material rather than the mask. Further, upon collision of the reactive ion with the substrate, a volatile product is produced that is exhausted out of the etching chamber thereby eliminating redeposition. Figure 7.22 shows the result of reactive ion etching on a glass waveguide using a mixture of C_2F_6 and C_2H_4 gases. Note the vertical walls and smooth sides.

For semiconductor waveguides, wet chemical etching provides a simple alternative for forming 3-D guides. Masking is performed in the standard fashion (see Figure 7.21) with either isotropic or anisotropic etchants used to remove exposed material. For isotropic etchants, material is removed at essentially an equal rate in all directions, whereas anisotropic etches attack specific crystallographic planes more rapidly than others. This can result in tapered waveguides such as those shown in Figure 7.23.

7.5.2 Raised-Channel Waveguides Formed by Shadow Masking

Shadow-mask fabrication refers to the process of depositing waveguiding films on a substrate through the openings in a mask located above but not in contact with the substrate. Because intimate contact with the substrate is avoided, the influence of roughness in the mask edges is eliminated. This technique has been used, for example, to fabricate tapered strip guides by the sputtering Ta_2O_5 through a slotted glass mask onto a glass substrate.

More recently, an extension of this technique using MBE has been developed that allows very precise patterning of a broad range of thin-film semiconductor structures for OIC applications. Figure 7.24 shows the basic arrangement used. A mechanical mask having fixed position is located slightly above a substrate on which the film pattern is to be written. Substrate motion is controlled in both the x and y axes. The mask is typically made from thin ($\sim$500-μm) silicon with square, rectangular, or

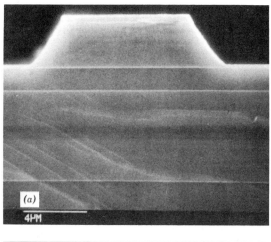

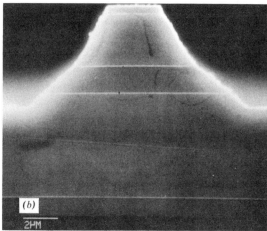

Figure 7.23 Wet chemically etched semiconductor waveguide on GaAs. (*a*) Strip loaded; (*b*) ridge. (After Walker and Goodfellow (Ref. 75). Reproduced by permission of the authors and the Institution of Electrical Engineers.)

circular holes etched through using standard photolithographic processes. Assume, for simplicity, that substrate motion is along one axis, say x. The thickness of the deposited film underneath the hole will be linearly proportional to the rate of epilayer growth at the surface and inversely proportional to the substrate velocity along x. Thus, for a fixed mask opening the thickness of the deposited film along x can be controlled by varying both beam flux and substrate velocity. Similarly, the film profile in the $\hat{y}$ direction can be controlled by adjusting the substrate velocity along y. By changing the beam composition with substrate position, the chemical and optical composition can be controlled in the plane of the film. Use of this method has been demonstrated in GaAs where strip waveguides only 0.3 μm wide have been fabricated. These structures exhibited mirrorlike sharply defined side walls.

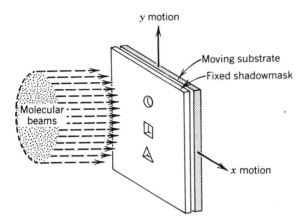

Figure 7.24 Method for defining 3-D guides using MBE by shadow masking. (After Tsang and Cho (Ref. 71). Reproduced by permission of the authors and the American Institute of Physics.)

7.5.3 Buried-Channel Waveguides

The diffusion, implantation, and ion-exchange processes can also be used in conjunction with a mask-making step to produce 3-D channel waveguides. The mask is made by first depositing a protective layer of photoresist, metal, or oxide onto the substrate surface. Using standard photolithographic procedures, the channel region of the waveguide is next defined by etching a slot or window into the protective layer. The interchange of atoms or ions between the substrate and surrounding environment through the mask window produces a guiding region within the substrate. The mask is subsequently removed leaving the channel waveguide shown in Figure 7.25. Channel waveguides offer a number of advantages over raised rib or strip guides. Because the diffused species are generally distributed in a smooth, continuous fashion, scattering losses of the type associated with rough wall surfaces in the strip or ridge guides are eliminated. Second, the fabrication process is somewhat simpler than for raised rib or strip guides. In addition the waveguide need only be fabricated over those regions of the substrate where it desired. Rib and strip 3-D structures must start with a 2-D guide fabricated over the entire substrate.

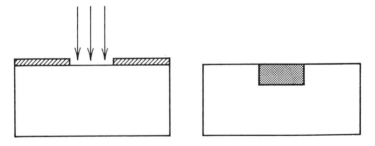

Figure 7.25 Method for producing 3-D channel waveguides using masking and atom or ion interchange with substrate.

7.6 OPTICAL-FIBER WAVEGUIDES FORMED BY FLAME HYDROLYSIS

Flame hydrolysis is a process used to form amorphous doped and undoped glass layers by the combustion of metal chloride vapors in an oxygen/hydrogen gas flame. The technique is primarily used to form low-loss optical fibers but has also recently been demonstrated as a means for deposition of very low loss SiO_2 and TiO_2 glass films on silicon substrates.

Pure silicon glass is formed by mixing $SiCl_4$ vapor with the fuel gases O_2 and CH_4 (methane) and combusting in the flame of a gas burner. The heat of the flame, along with the oxygen, causes the $SiCl_4$ vapors to react forming very small silica glass spheres (~ 0.1 μm diam) called "soot" which are directed toward a target. The basic reaction is

$$SiCl_4(v) + O_2(v) \longrightarrow SiO_2(s) + 2Cl_2(v)$$

Water is also given off as a product of the combustion. The properties of the glass can be modified by the addition of small amounts of metal chloride vapors such as $GeCl_4$, $POCl_3$, and BCl_3 which react with O_2 in an analogous fashion to $SiCl_4$ forming the glasses GeO_2, P_2O_5, and B_2O_3. These dopant glasses have two significant effects. They allow the permittivity of the composite glass to be varied and they lower the melting point relative to pure SiO_2 glass.

In the outside vapor deposition or OVD process, the hot soot stream from the burner is directed at a rotating target or "bait" rod made of graphite, fused silica, or a crystalline ceramic. The glass soot sticks to the rod and layer by layer a porous

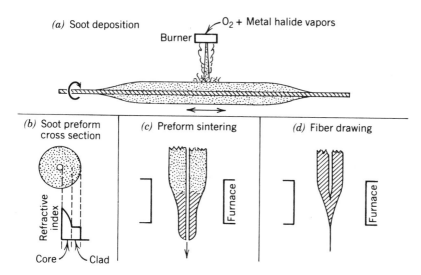

Figure 7.26 Steps in the fabrication of an optical fiber using the OVD process. (A. H. Cherin, *An Introduction to Optical Fibers*. McGraw-Hill, New York. Copyright © 1983, Bell Laboratories, Inc.)

cylindrical preform is built up as shown in Figure 7.26a. By varying the flow rates of the various doping vapors, the permittivity variation in the radial direction can be controlled as shown in Figure 7.26b. After completion of the soot deposition, the target rod is removed from the preform. The preform is then heated in a zone furnace to approximately 1500°C in a He atmosphere containing a few percent Cl_2. This process, called sintering, results in a dense bubble-free clear glass preform having a small hole along the center, as shown in Figure 7.26c. The addition of Cl_2 during the sintering process has the effect of purging OH^- ions bound to the glass by reacting with them to form HCl and O_2. Removal of OH^- ions is of paramount importance since they cause optical absorption in the wavelength region of principle interest for integrated optics.

The final fiber is obtained using a drawing process in which the preform is slowly fed into a furnace heated to a temperature between 1800 and 2000°C and the melted end is pulled out at high speed, as shown in Figure 7.26d.

A similar process using a combination of $SiCl_4$ and $TiCl_4$ vapors and BCl_3 and PCl_3 dopant gases has also been used in fabricating planar step dielectric guides on silica glass substrates. A very low propagation loss of 0.15 dB/cm at 0.63 μm was obtained.

PROBLEMS

7.1 A waveguide is to be fabricated using a Corning 7059 glass substrate. Which of the following glasses would yield the highest number of waveguide modes if used for the same thickness guiding layer?

(a) Barium silicate
(b) GeO_2
(c) Nb_2O_5
(d) Corning 7059

7.2 What are the merits of magnetron sputtering relative to glow discharge sputtering?

7.3 It is found that in sputtering an SiO_2 target in an Ar atmosphere, the resulting deposited SiO_2 film is Si rich. What could be done to improve the stoichiometry?

7.4 It is determined that when a Nb_2O_5 target is sputtered, the composition of the film when deposited on Corning 7059 glass is somewhat different than that obtained on a fused quartz substrate under identical sputtering conditions. Give possible explanations.

7.5 What is the principal advantage offered by polymer-film waveguides over sputtered-film ones?

7.6 Estimate the number of TE modes that can be supported by a waveguide formed by a 30-min Ag ion migration into a soda-lime glass substrate under an applied electric field of 100 V at a temperature of 250°C. Assume $\lambda_0 = 0.63$ μm.

7.7 Explain why the addition of a salt such as $LiNO_3$ to benzoic acid reduces the amount of proton exchange in a $LiNbO_3$ waveguide.

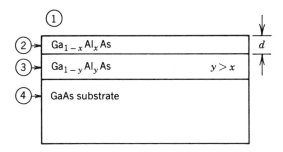

Figure 7.27 Waveguide geometry for problem 7.8.

7.8 Figure 7.27 shows a proposed dielectric waveguide. Let us define the difference in mole fraction of Al between regions 3 and 2 as $\Delta = y - x$.

For small Δ, the difference in refractive index between regions 2 and 3 is found to be empirically given by

$$n_2 - n_3 \cong 0.4\Delta$$

Use the results of problem 4.7 to show that for single-mode propagation we require

$$\frac{1.25}{n}\left(\frac{\lambda_0}{4d}\right)^2 < \Delta < \frac{11.25}{n}\left(\frac{\lambda_0}{4d}\right)^2$$

7.9 With reference to Figure 7.28, the refractive-index increase $\Delta n(x)$ for proton exchange in a Z-cut LiNbO$_3$ substrate (crystallographic Z axis is along the crystal plate normal) is found to be represented to good approximation by a step function as

$$\Delta n(x) = 0.12u_{-1}(x + d)$$

where

$$u_{-1}(x + d) = 1 \quad -d < x < 0$$
$$= 0 \quad x \quad < -d$$

The diffusion distance d is a function both of diffusion time t and diffusion temperature T. Assume that d can be approximated by

$$d = 2\sqrt{t \times D(T)}$$

where the diffusion coefficient D is

$$D(T) = D_0 \exp(-T_0/T)$$

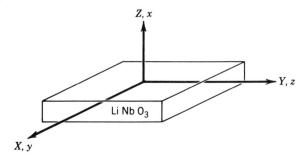

Figure 7.28 Crystallographic orientation for Z-cut LiNbO$_3$.

and

$$T_0 = 1.13 \times 10^4 \text{ K}°$$

$$D_0 = 5.1 \times 10^5 \text{ μm}^2/\text{s}$$

(a) Based upon the discussion in Section 7.3.2, will the proton exchange waveguide support wave propagation along the crystallographic X axis? If so, TE or TM modes or both? Repeat for wave propagation along the crystallographic Y axis.

(b) At a temperature $T = 200°C$ how long must the substrate ($n = 2.20$) be heated for the resulting guide to have the fundamental mode just at cutoff? Assuming $\lambda_0 = 0.6328$ μm?

(c) Repeat part (b) for $T = 250°C$.

7.10 For X-, Y-, and Z-cut plates (crystallographic X, Y, and Z, axes normal to substrate, respectively) describe, based upon the discussion of Section 7.3.2, which propagation directions and polarizations will support guided modes for proton exchanged LiTaO$_3$ and LiNbO$_3$ waveguides.

7.11 (a) Assume a slab semiconductor waveguide is fabricated by depositing a GaAs film of thickness d having an electron carrier concentration N_f on top of a GaAs substrate having carrier concentration N_s. Using the results of Example 2.3 and problem 4.7, show that for guidance to occur

$$N_s - N_f > \frac{(\pi c)^2 \, m^* \, \epsilon_0}{(2qd)^2}$$

where m^* is the effective electron mass.

(b) Compute $N_s - N_f$ for GaAs assuming that $m^* \cong 0.068 \, m$, where m is the free electron mass and $d = 3$ μm.

7.12 The refractive index in Ga$_{1-x}$Al$_x$As can be modeled empirically using the so-called Sellmeier equation as

$$n^2(x) = A(x) + \frac{B}{\lambda_0^2 - C(x)} - D(x)\lambda_0^2$$

where λ_0 is in micrometers and

$$A(x) = 10.906 - 2.92x$$
$$B = 0.97501$$
$$C(x) = (0.52886 - 0.735x)^2 \quad x \leq 0.36$$
$$= (0.30386 - 0.105)^2 \quad x > 0.36$$
$$D(x) = 0.002467 (1.41x + 1)$$

(a) What is the value for the effective index for the waveguide of Figure 7.11 for $\lambda_0 = 0.9$ μm?

(b) What is the $1/e$ distance that the fields extend beyond the core?

7.13 Figure 7.8 shows the variation of implant depth with He^+ ion-implant energy for a dose of $6 \times 10^{15}/cm^2$. Use Figure 7.8a to estimate approximately the required implant energy necessary to fabricate a single-mode (fundamental just above cutoff) guide at the He–Ne laser wavelength. The refractive index of fused silica is 1.45.

7.14 Show that the total dose N_d given in Eq. 7.3 is related to the concentration $C(x)$ by

$$N_d = \int_{-\infty}^{\infty} C(x)\, dx$$

7.15 Discuss the relative merits and disadvantages of LPE, CVD, and MBE film deposition techniques. What is the primary advantage of 3-D guides formed using MBE shadow masking? Are there any disadvantages to this method? What is the purpose of adding HCl to the sintering process in the formation of optical fibers? What role do the metal chloride vapors $GeCl_4$, $POCl_3$, and BCl_3 play in the fiber formation?

REFERENCES

1. Aagard, R. L. "Optical waveguide characteristics of reactive dc-sputtered niobium pentoxide films." *Applied Physics Letters* 27 (1975): 605–607.

2. Becker, R. A. "Comparison of guided-wave interferometric modulators fabricated on $LiNbO_3$ via Ti indiffusion and proton exchange." *Applied Physics Letters* (1983): 131–133.

3. Blum, F. A., Shaw, D. W., and Holton, W. C. "Optical striplines for integrated optical circuits in epitaxial GaAs." *Applied Physics Letters* 25 (1974): 116–118.

4. Bornholdt, C., et al. "Passive optical GaInAsP/InP waveguides. *Electronics Letters* 19 (1983): 81–82.

5. Burns, W. K., Bulmer, C. H., and West, E. J. "Application of Li_2O compensation techniques to Ti-diffused $LiNbO_3$ planar and channel waveguides." *Applied Physics Letters* 33 (1978): 70–72.

6. Chandra, P., Coldren, L. A., and Strege, K. E. "Refractive index data from $Ga_xIn_{1-x}As_yP_{1-y}$ films." *Electronics Letters* 17 (1981): 6–7.

7. L. L. Chang, "Molecular beam epitaxy." In *Handbook on Semiconductors*. Vol. 3. Edited by S. P. Keller. New York: North-Holland, 1980.

8. Chartier, G. H., et al. "Graded-index surface or buried waveguides by ion exchange in glass." *Applied Optics* 19 (1980): 1092–1095.

9. Chartier, G. H., et al. "Fast fabrication method for thick and highly multimode optical waveguides." *Electronics Letters* 13 (1977): 763–764.

10. Chartier, G. H., et al. "Optical waveguides fabricated by electric-field controlled ion exchange in glass." *Electronics Letters* 14 (1978): 132–134.

11. Cho, A. Y., and Arthur, J. R. "Molecular beam epitaxy." In *Progress in Solid-State Chemistry*. Vol. 10. New York: Pergamon, 1975.

12. Clark, D. F., et al. "Characterization of proton-exchange slab optical waveguides in Z-cut $LiNbO_3$. *Journal of Applied Physics* 54 (1983): 6218–6220.

13. Destefanis, G. L., Townsend, P. D., and Gailliard, J. P. "Optical waveguides in $LiNbO_3$ formed by ion implantation of helium." *Applied Physics Letters* 32 (1978): 293–294.

14. Destefanis, G. L., et al. "Formation of waveguides and modulators in $LiNbO_3$ by ion implantation." *Journal of Applied Physics* 50 (1979): 7898–7905.

15. Dupuis, R. D., and Dapkus, P. D. "Preparation and properties of $Ga_{1-x}Al_xAs$–GaAs heterostructure lasers grown by metalorganic chemical vapor deposition." *IEEE Journal of Quantum Electronics* QE-15 (1979): 128–135.

16. EerNisse, E. P., and Norris, C. B. "Introduction rates and annealing of defects in ion-implanted SiO_2 layers on Si." *Journal of Applied Physics* 45 (1974): 5196–5205.

17. Fukuda, T., and Hirano, H. "Capillary liquid epitaxial growth of $LiNbO_3$ and $LiTaO_3$ single-crystal thin films." *Applied Physics Letters* 28 (1976): 575–577.

18. Furuta, H., Noda, H., and Ihaya, A. "Novel optical waveguide for integrated optics." *Applied Optics* 13 (1974): 322–326.

19. Garmire, E., Lovelace, D. F., and Thompson, G. H. B. "Diffused two-dimensional optical waveguides in GaAs." *Applied Physics Letters* 26 (1975): 329–331.

20. Garmire, E., Stoll, H., and Yariv, A. "Optical waveguiding in proton-implanted GaAs." *Applied Physics Letters* 21 (1972): 87–88.

21. Garvin, H. L., et al. "Ion beam micromachining of integrated optics components." *Applied Optics* 12 (1973): 455–459.

22. Goell, J. E. "Electron-resist fabrication of bends and couplers for integrated optical circuits." *Applied Optics* 12 (1973): 729–736.

23. Goell, J. E. "Rib waveguide for integrated optical circuits." *Applied Optics* 12 (1973): 2797–2798.

24. Goell, J. E., and Standley, R. D. "Ion bombardment fabrication of optical waveguides using electron resist masks." *Applied Physics Letters* 21 (1972): 72–73.

25. Goodwin, M., and Stewart, C. "Proton-exchanged optical waveguides in *Y*-cut lithium niobate." *Electronics Letters* 19 (1983): 223–224.

26. Griffiths, G. J., and Esdaile, R. J. "Analysis of titanium diffused planar optical waveguides in lithium niobate." *IEEE Journal of Quantum Electronics* QE-20 (1984): 149–159.

27. Hafich, M., Chen, D., and Huber, J. "Properties of optical waveguides formed by thermal migration of thallium ions in glass." *Applied Physics Letters* 33 (1978): 997–999.

28. Hammer, J. M., and Phillips, W. "Low-loss single-mode optical waveguides and efficient high-speed modulators of $LiNb_xTa_{1-x}O_3$." *Applied Physics Letters* 24 (1974): 545–547.

29. Heibei, J., and Voges, E. "Refractive index profiles of ion-implanted fused silica." *Physica Status Solidi* 57 (1980): 609–618.

30. Hensler, D. H., *et al.* "Optical propagation in sheet and pattern generated films of Ta_2O_5." *Applied Optics* 10 (1971): 1037–1042.

31. Houghton, A. J. N., *et al.* "Low-loss optical waveguides in MBE-grown GaAs/GaAlAs heterostructures." *Optics Communications* 46 (1983): 164–166.

32. Hsieh, J. J., "Liquid-phase epitaxy." In *Handbook on Semiconductors.* Vol. 3. Edited by S. P. Keller. New York: North-Holland, 1980.

33. Hunsperger, R. G. *Integrated Optics: Theory and Technology.* New York: Springer-Verlag, 1982.

34. Izawa, T., Mori, H., Murakami, Y., and Shimizu, N. "Deposited silica waveguide for integrated optical circuits." *Applied Physics Letters* 38 (1981): 483–485.

35. Izawa, T., and Nakagome, H. "Optical waveguide formed by electrically induced migration of ions in glass plates." *Applied Physics Letters* 21 (1972): 57–59.

36. Jackel, J. L., *et al.* "Reactive ion etching of $LiNbO_3$." *Applied Physics Letters* 38 (1981): 907–909.

37. Jackel, J. L., Ramaswamy, V., and Lyman, S. P. "Elimination of out-diffused surface guiding in titanium-diffused $LiNbO_3$." *Applied Physics Letters* 38 (1981): 509–511.

38. Jackel, J. L., Rice, C. E., and Veselka, J. J. "Proton exchange for high-index waveguides in $LiNbO_3$." *Applied Physics Letters* 41 (1982): 607–608.

39. Jackel, J. L., Rice, C. E., and Veselka, J. J. "Composition control in proton-exchanged $LiNbO_3$." *Electronics Letters* 19 (1983): 387–388.

40. Kaminow, I. P., and Carruthers, J. R. "Optical waveguiding layers in $LiNbO_3$ and $LiTaO_3$." *Applied Physics Letters* 22 (1973): 326–328.

41. Kaminow, I. P., *et al.* "Lithium niobate ridge waveguide modulator." *Applied Physics Letters* 24 (1974): 622–624.

42. Kawachi, M., Yasu, M., and Edahiro, T. "Fabrication of Sio_2–TiO_2 glass planar optical waveguides by flame hydrolysis deposition." *Electronics Letters* 19 (1983): 583–584.

43. Kressel, H., and Nelson, H. "Properties and applications of III-V compound films deposited by liquid phase epitaxy." In *Physics of Thin Films.* Vol. 7. Edited by G. Hass, M. H. Francombe, and R. W. Hoffman. New York: Academic, 1973.

44. Lee, Y., and Wang, S. "Tantalum oxide light guide on lithium tantalate." *Applied Physics Letters* 25 (1974): 164–166.

45. Leonberger, F. J., Donnelly, J. P., and Bozler, C. O. "Low-loss GaAs $p^+n^-n^+$ three-dimensional optical waveguides." *Applied Physics Letters* 28 (1976): 616–619.

46. Manasevit, H. M., and Simpson, W. I. "The use of metal-organics in the preparation of semiconductor materials." *Solid State Science* 116 (1969): 1725–1732.

47. Mayer, J. W., Eriksson, L., and Davies, J. A. *Ion Implantation in Semiconductors.* New York: Academic, 1971.

48. Miyazawa, S. "Growth of LiNbO₃ single-crystal film for optical waveguides." *Applied Physics Letters* 23 (1973): 198–200.

49. Naik, I. K. "Low-loss integrated optical waveguides fabricated by nitrogen ion implantation." *Applied Physics Letters* 43 (1983): 519–520.

50. Nelson, H. "Epitaxial growth from the liquid state and its application to the fabrication of tunnel and laser diodes." *RCA Review* (Dec. 1963): 603–615.

51. Noda, J., *et al.* "Strip-loaded waveguide formed in a graded-index LiNbO₃ planar waveguide." *Applied Optics,* 17 (1978): 1953–1958.

52. Ohke, S., *et al.* "Optical waveguide in GaAs by drive-in diffusion technique." *Japanese Journal of Applied Physics* 22 (1983): 1106–1108.

53. Ostrowsky, D. B., and Vanneste, C. "Thin films for integrated optics." In *Physics of Thin Films.* Vol. 10. Edited by G. Hass and M. H. Francombe. New York: Academic, 1978.

54. Pearsall, T. P. *GaInAsP Alloy Semiconductors.* New York: Wiley, 1982.

55. Pelosi, P. M., *et al.* "Propagation characteristics of trapezoidal cross-section ridge optical waveguides: An experimental and theoretical investigation." *Applied Optics* 17 (1978): 1187–1193.

56. Ramaswamy, V., and Weber, H. P. "Low-loss polymer films with adjustable refractive index." *Applied Optics* 12 (1973): 1581–1583.

57. Rand, M. J., and Standley, R. D. "Silicon oxynitride films on fused silica for optical waveguides." *Applied Optics* 11 (1972): 2482–2488.

58. Ranganath, T. R., Tsang, W. T., and Wang, S. "Tapered edge ridge waveguides for integrated optics." *Applied Optics* 14 (1975): 1847–1853.

59. Sasaki, H., Kushibiki, J., and Chubachi, N. "Efficient acousto-optic TE-TM mode conversion in ZnO films." *Applied Physics Letters* 25 (1974): 476–477.

60. Schineller, E. R., Flam, R. P., and Wilmot, D. W. "Optical waveguides formed by proton irradiation of fused silica." *Journal of the Optical Society of America* 58 (1968): 1171–1175.

61. Shteingart, L. M. "Fabrication of optical waveguides in LiNbO₃ and LiTaO₃ crystals by ion irradiation." *Sov. Physics Tehnical Physics* 27 (1982): 1414–1415.

62. Somekh, S., *et al.* "Channel optical waveguides and directional couplers in GaAs-imbedded and ridged." *Applied Optics* 13 (1974): 327–330.

63. Sosnowski, T. P., and Weber, H. P. "Thin birefringent polymer films for integrated optics." *Applied Physics Letters* 21 (1972): 310–311.

64. Standley, R. D., Gibson, W. M., and Rodgers, J. W. "Properties of ion-bombarded fused quartz for integrated optics." *Applied Optics* 11 (1972): 1313–1316.

65. Standley, R. D., and Ramaswamy, V. "Nb-diffused LiTaO₃ optical waveguides: Planar and embedded." *Applied Physics Letters* 25 (1974): 711–713.

66. Stewart, G., *et al.* "Planar optical waveguides formed by silver-ion migration in glass." *IEEE Journal of Quantum Electronics* QE-13 (1977): 192–200.

67. Stulz, L. W. "Titanium in-diffused LiNbO₃ optical waveguide fabrication." *Applied Optics* 18 (1979): 2041–2044.

68. Takada, S., *et al.* "Optical waveguides of single-crystal LiNbO₃ film deposited by rf sputtering." *Applied Physics Letters* 24 (1974): 490–492.

69. Tien, P. K., Smolinsky, G., and Martin, R. J. "Thin organosilicon films for integrated optics." *Applied Optics* 11 (1972): 637–642.

70. Tsang, W. T. "The preparation of GaAs thin-film optical components by molecular beam epitaxy using Si shadow masking technique." *Applied Physics Letters* 35 (1979): 792–794.

71. Tsang, W. T., and Cho, A. Y. "Molecular beam epitaxial writing of patterned GaAs epilayer structures." *Applied Physics Letters* 32 (1978): 491–493.

72. Ulrich, R., and Weber, H. P. "Solution-deposited thin films as passive and active light-guides." *Applied Optics* 11 (1972): 428–434.

73. Viljanen, J., and Leppihalme, M. "Fabrication of optical strip waveguides with nearly circular cross section by silver ion migration technique." *Journal of Applied Physics* 51 (1980): 3563–3565.

74. Vossen, J. L., and Kern, W. *Thin Film Processes.* New York: Academic, 1978.

75. Walker, R. G., and Goodfellow, R. C. "Attenuation measurements on MOCVD grown GaAs/GaAlAs optical waveguides." In Proceedings, IEE Second European Conference on Integrated Optics, Firenze, Italy, October 1983. 61–64.

76. Walker, R. G., Wilkinson, C. D. W., and Wilkinson, J. A. H. "Integrated optical waveguiding structures made by silver ion-exchange in glass. 1. The propagation characteristics of stripe ion-exchanged waveguides; A theoretical and experimental investigation." *Applied Optics* 22 (1983): 1923–1936.

77. Weber, H. W., *et al.* "Light-guiding structures of photoresist films." *Applied Physics Letters* 20 (1972): 143–145.

78. Wei, D. T. Y., Lee, W. W., and Bloom, L. R. "Quartz optical waveguide by ion implantation." *Applied Physics Letters* 22 (1973): 5–7.

79. Wenzlik, K., Heibei, J., and Voges, E. "Refractive index profiles of helium implanted LiNbO₃ and LiTaO₃." *Physica Status Solidi* 61 (1980): K207–K211.

80. Yin, Z., and Garside, B. K. "Low-loss GeO₂ optical waveguide fabrication using low deposition rate rf sputtering." *Applied Optics* 21 (1982): 4324–4328.

81. Yang, E. S. *Fundamentals of Semiconductor Devices.* New York: McGraw-Hill, 1978.

82. Yi-Yan, A. "Index instabilities in proton-exchanged LiNbO₃ waveguides." *Applied Physics Letters* 42 (1983): 633–635.

83. Zelmon, D. E. "A low-scattering graded-index SiO₂ planar optical waveguide thermally grown on silicon." *Applied Physics Letters* 42 (1983): 565–566.

CHAPTER 8

Mode Coupling

8.1 INTRODUCTION

All of the guiding structures discussed in this text are representatives of a broad class known as cylindrical dielectric waveguides. By definition, a cylindrical dielectric waveguide is a structure whose permittivity is invariant along the propagation direction z as shown in Figure 8.1. Assuming a free-space permeability μ_0, then the waveguide geometry is completely defined by specifying the variation of the permittivity ϵ with x and y.

No matter how complicated the function $\epsilon(x, y)$ may be, or equivalently, regardless of how bizarre the shape of the cylindrical dielectric waveguide, several important general statements can be made concerning the nature of the guided modes supported by the structure. These relationships, derived in this chapter, are important for the subsequent analysis of coupled-mode theory.

8.2 WAVEGUIDE SYMMETRY PROPERTIES

8.2.1 z-Reversal Symmetry

Let us assume that the dielectric waveguide geometry has been defined by specifying a particular form for $\epsilon(x, y)$. We further assume that the field solutions for a particular forward-travelling mode have been obtained in some fashion. These fields are of the form

$$\left. \begin{array}{l} \mathbf{E}(x, y, z) = \mathcal{E}(x, y) \\ \mathbf{H}(x, y, z) = \mathcal{H}(x, y) \end{array} \right\} e^{-jk_z z} \tag{8.1}$$

and represent the solution to Maxwell's curl equations:

$$\nabla \times (\mathcal{E}e^{-jk_z z}) = -j\omega\mu(\mathcal{H}e^{-jk_z z})$$
$$\nabla \times (\mathcal{H}e^{-jk_z z}) = +j\omega\epsilon(\mathcal{E}e^{-jk_z z}) \tag{8.2}$$

We can break up the ∇ operator into longitudinal and transverse components of the form

$$\nabla = \nabla_t + \hat{z} \, \partial/\partial z$$

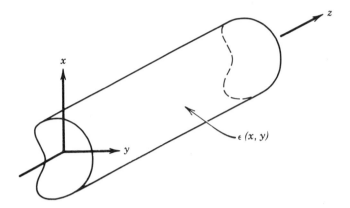

Figure 8.1 Generalized cylindrical waveguide.

where

$$\nabla_t \equiv \hat{x}\,\frac{\partial}{\partial x} + \hat{y}\,\frac{\partial}{\partial y}$$

The fields $\mathscr{E}$ and $\mathscr{H}$, can also be separated into longitudinal and transverse components:

$$\mathscr{E} = \mathscr{E}_t + \hat{z}\mathscr{E}_z$$

$$\mathscr{E}_t \equiv \hat{x}\mathscr{E}_x + \hat{y}\mathscr{E}_y$$

with $\mathscr{H}$ similarly separated. If the fields and operators are substituted into the first of Maxwell's curl equations we obtain

$$(\nabla_t + \hat{z}\,\partial/\partial z) \times (\mathscr{E}_t + \hat{z}\mathscr{E}_z)e^{-jk_z z} = -j\omega\mu(\mathscr{H}_t + \hat{z}\mathscr{H}_z)e^{-jk_z z}$$

Expanding the left side of the above expression yields

$$\nabla_t \times \mathscr{E}_t e^{-jk_z z} + \nabla_t \times \hat{z}\mathscr{E}_z e^{-jk_z z} + \hat{z} \times \mathscr{E}_t(\partial/\partial z\ e^{-jk_z z}) + \hat{z} \times \hat{z}\mathscr{E}_z(\partial/\partial z\ e^{-jk_z z})$$
$$= -j\omega\mu(\mathscr{H}_t + \hat{z}\mathscr{H}_z)e^{-jk_z z}$$

Note that the only nonzero terms in the first cross-product are along $\hat{z}$. Similarly, the second and third terms contain only $\hat{x}$ and $\hat{y}$ vector cross-products with $\hat{z}$ and therefore have no component in the $\hat{z}$ direction. The fourth term is identically zero. Thus, equating longitudinal and transversal components on both sides of this expression yields Eqs. 8.3 and 8.4. Performing a similar set of operations on the curl equation for $\mathscr{H}$ yields

Eqs. 8.5 and 8.6.

$$\nabla_t \times (\mathcal{E}_t) = -j\omega\mu\hat{z}(\mathcal{H}_z) \tag{8.3}$$

$$\nabla_t \times \hat{z}(\mathcal{E}_z) - jk_z\hat{z} \times (\mathcal{E}_t) = -j\omega\mu(\mathcal{H}_t) \tag{8.4}$$

$$\nabla_t \times (\mathcal{H}_t) = +j\omega\epsilon\hat{z}(\mathcal{E}_z) \tag{8.5}$$

$$\nabla_t \times \hat{z}(\mathcal{H}_z) - jk_z\hat{z} \times (\mathcal{H}_t) = +j\omega\epsilon(\mathcal{E}_t) \tag{8.6}$$

Thus, the field quantities in parentheses above represent the solutions to Maxwell's equations for the waveguide defined by the $\epsilon(x, y)$ indicated in these equations.

It is possible to show that if the fields above are a solution to Maxwell's equations, and thus to the waveguide geometry, a second related solution also exists. If the following replacement is made in Eqs. 8.3–8.6

$$\mathcal{E}_t \longrightarrow \mathcal{E}_t \qquad \mathcal{H}_t \longrightarrow -\mathcal{H}_t$$

$$\mathcal{E}_z \longrightarrow -\mathcal{E}_z \qquad \mathcal{H}_z \longrightarrow \mathcal{H}_z \tag{8.7}$$

$$k_z \longrightarrow -k_z$$

we obtain the identical set of equations. Thus, if the fields on the left side of the arrows are a solution to the waveguide problem described by Maxwell's equations then the fields on the right side of the arrows are also a solution. Note that the second solution corresponds to a wave traveling in the negative $\hat{z}$ direction with the same phase velocity as the corresponding forward-traveling wave. Therefore, on cylindrical dielectric waveguides we conclude that for each forward-traveling mode having propagation constant k_z there corresponds a backward-traveling mode with propagation constant $-k_z$ and the fields of the two modes are related by Eq. 8.7.

8.2.2 Time-Reversal Symmetry

Suppose we take the complex conjugate of Maxwell's equations 8.2, thereby obtaining

$$\nabla \times (\mathcal{E}^* e^{+jk_z z}) = -j\omega\mu(-\mathcal{H}^* e^{+jk_z z})$$
$$\nabla \times (-\mathcal{H}^* e^{+jk_z z}) = j\omega\epsilon(\mathcal{E}^* e^{+jk_z z}) \tag{8.8}$$

The resultant equations are identical in form with Eq. 8.2 and therefore we conclude that another solution is

$$\left. \begin{array}{l} \mathbf{E} = \mathcal{E}^* \\ \mathbf{H} = -\mathcal{H}^* \end{array} \right\} e^{+jk_z z}$$

Thus, the above set of fields must be an alternative but equivalent representation for our backward-traveling mode. Comparing this result with the field representation for the backward mode given in the last section, we have the relations

$$\mathcal{E}_t = \mathcal{E}_t^* \qquad \mathcal{E}_z = -\mathcal{E}_z^*$$

$$\mathcal{H}_t = \mathcal{H}_t^* \qquad \mathcal{H}_z = -\mathcal{H}_z^*$$

This implies that the transverse fields can be chosen to be purely real and the longitudinal fields purely imaginary.

8.3   LORENTZ RECIPROCITY THEOREM

Often it is necessary to analyze the optical properties of guiding dielectric structures that are too complicated to solve conveniently by either the ray approach or by a rigorous application of Maxwell's equations with appropriate boundary conditions. However, many of these problems can be considered to be small perturbations of a simpler problem which we already know how to solve. A powerful mathematical technique that is applicable to problems of this nature is coupled-mode theory. In this method we try to represent the field solutions on our complicated or perturbed waveguide in terms of a superposition of simpler, known solutions of the unperturbed guide. For example, our unperturbed waveguide might be the slab dielectric structure shown in Figure 8.2. Typical perturbed waveguides that we might wish to analyze might consist of any of the structures shown in Figure 8.3. To obtain the relationship between the perturbed and unperturbed field quantities we derive a very powerful theorem known as the Lorentz reciprocity theorem.

8.3.1   Formulation

Let **E** and **H** represent any allowable solution to Maxwell's equations on the unperturbed waveguide, which is assumed to be lossless and characterized by permittivity $\epsilon(x, y, z)$. As can be seen from Figure 8.3, the geometry of the perturbed waveguide is completely specified by its dielectric constant $\epsilon'(x, y, z)$. Let the perturbed fields associated with this structure be designated by **E′** and **H′**. Maxwell's equations for the two structures are therefore given by

$$\text{(a)} \ \nabla \times \mathbf{E} = -j\omega\mu\mathbf{H} \qquad \text{(c)} \ \nabla \times \mathbf{E}' = -j\omega\mu\mathbf{H}'$$

$$\text{(b)} \ \nabla \times \mathbf{H} = j\omega\epsilon\mathbf{E} \qquad \text{(d)} \ \nabla \times \mathbf{H}' = j\omega\epsilon'\mathbf{E}'$$

If we take the quantity

$$\mathbf{E}^* \cdot \text{(d)} - \mathbf{H}' \cdot \text{(a)}^* - \mathbf{H}^* \cdot \text{(c)} + \mathbf{E}' \cdot \text{(b)}^*$$

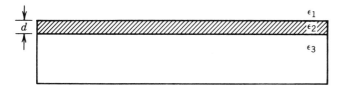

Figure 8.2 Unperturbed slab waveguide.

we obtain

$$(\mathbf{E}^* \cdot \nabla \times \mathbf{H}' - \mathbf{H}' \cdot \nabla \times \mathbf{E}^*) - (\mathbf{H}^* \cdot \nabla \times \mathbf{E}' - \mathbf{E}' \cdot \nabla \times \mathbf{H}^*)$$

$$= +j\omega(\epsilon' - \epsilon)\,\mathbf{E}^* \cdot \mathbf{E}'$$

Making use of the vector identity

$$\nabla \cdot (\mathbf{A} \times \mathbf{B}) = \mathbf{B} \cdot \nabla \times \mathbf{A} - \mathbf{A} \cdot \nabla \times \mathbf{B}$$

the above relationship simplifies to

$$\nabla \cdot [\mathbf{E}^* \times \mathbf{H}' + \mathbf{E}' \times \mathbf{H}^*] = -j\omega(\epsilon' - \epsilon)\,\mathbf{E}^* \cdot \mathbf{E}'$$

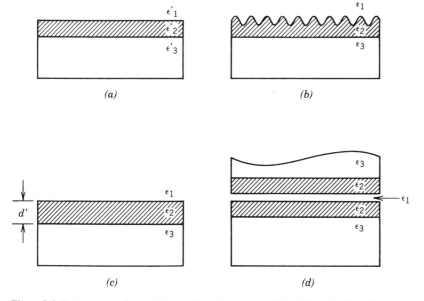

Figure 8.3 Various types of perturbations of the original waveguide. (*a*) Perturbed permittivity; (*b*) perturbed surface; (*c*) perturbed thickness; (*d*) introduction of second guide.

Integrating both sides of this equation over a volume V and making use of Gauss' theorem,

$$\iiint_V dV\, \nabla \cdot \mathbf{A} = \oiint_S \mathbf{A} \cdot d\mathbf{S}$$

yields

$$\oiint_S (\mathbf{E}^* \times \mathbf{H}' + \mathbf{E}' \times \mathbf{H}^*) \cdot d\mathbf{S} = -j\omega \iiint_V (\epsilon' - \epsilon)\, \mathbf{E}^* \cdot \mathbf{E}'\, dV \quad (8.9)$$

where the closed surface S surrounds the volume V. The above relation is known as the Lorentz reciprocity theorem. It lies at the heart of the solution of coupled-mode analysis.

For cylindrical guiding structures, we choose for the surface S a circular cylinder of infinite radius centered about the axis of the waveguide and infinitesimal width Δz along z as shown in Figure 8.4. Thus the integral over S can be broken up into three components:

$$\oiint_S = \iint_{S_{\mathrm{I}}} + \iint_{S_{\mathrm{II}}} + \iint_C$$

Since the guided-wave solutions decay exponentially away from the core or guiding region, the contribution from the surface C which is infinitely far from the waveguide, is zero. Let us further define

$$\mathbf{E}^* \times \mathbf{H}' + \mathbf{E}' \times \mathbf{H}^* = \mathbf{F}(x, y, z)$$

Then

$$\iint \mathbf{F}(x, y, z) \cdot d\mathbf{S} = -\iint_{S_{\mathrm{I}}} F_z(x, y, z)\, dS + \iint_{S_{\mathrm{II}}} F_z(x, y, z)\, dS$$

$$= -\iint_S F_z(x, y, z)\, dS + \iint_S F_z(x, y, z + \Delta z)\, dS$$

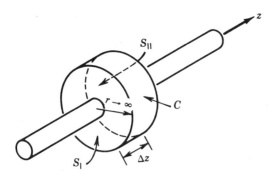

Figure 8.4 Surface over which Lorentz reciprocity theorem is to be evaluated.

We now take the limit of both sides of Eq. 8.9 as Δz approaches zero. In this limit, the fields contained in volume V are independent of z so that the volume integral on the right side of Eq. 8.9 reduces to a surface integral times the length Δz. Thus,

$$\lim_{\Delta z \to 0} \left\{ \iint_S [F_z(x, y, z + \Delta z) - F_z(x, y, z)] \, dS = -j\omega \Delta z \iint_S (\epsilon' - \epsilon) \, \mathbf{E}' \cdot \mathbf{E}^* \, dS \right\}$$

Dividing both sides by Δz then yields

$$\iint_S \frac{\partial}{\partial z} F_z(x, y, z) \, dS = -j\omega \iint_S (\epsilon' - \epsilon) \, \mathbf{E}' \cdot \mathbf{E}^* \, dS$$

One additional simplification can be made by noting that F_z consists of the $\hat{z}$ components of cross-products of the form $\mathbf{A} \times \mathbf{B}$. Representing these fields in terms of transverse and longitudinal components

$$\mathbf{A} = \mathbf{A}_t + \hat{z} A_z$$

$$\mathbf{B} = \mathbf{B}_t + \hat{z} B_z$$

then the $\hat{z}$ component of $\mathbf{A} \times \mathbf{B}$ is found by direct substitution to be

$$(\mathbf{A} \times \mathbf{B})_z = \mathbf{A}_t \times \mathbf{B}_t$$

We have therefore as the Lorentz reciprocity theorem for the general class of cylindrical waveguides

$$\iint_S \frac{\partial}{\partial z}(\mathbf{E}_t^* \times \mathbf{H}_t' + \mathbf{E}_t' \times \mathbf{H}_t^*) \cdot \hat{z} \, dS = -j\omega \iint_S (\epsilon' - \epsilon) \, \mathbf{E}' \cdot \mathbf{E}^* \, dS \quad (8.10)$$

8.4 MODE ORTHOGONALITY RELATIONS

To obtain expressions for the perturbed fields in terms of the unperturbed ones, it is necessary to obtain orthogonality relations for the allowed modes that can exist on the unperturbed structure. The reader has used such relations, perhaps unknowingly, in an elementary study of Fourier series. For example, consider some periodic voltage wave train $v(t)$ with periodicity T which for simplicity is assumed to be an odd function of time. The function $v(t)$ can be represented as a superposition of harmonics:

$$v(t) = \sum_n a_n \sin n\omega t \quad (8.11)$$

Thus we describe a complicated function, $v(t)$, in terms of a superposition of simpler functions, harmonics of sines.

If the amplitude coefficients a_n are known, then $v(t)$ is completely described. These coefficients can be obtained because of our knowledge of the orthogonality relationship between various harmonics of the sine functions:

$$\frac{2}{T} \int_0^T dt\{\sin(n\omega t)\ \sin(m\omega t)\} = \delta_{n,m} \tag{8.12}$$

where the Kronecher delta $\delta_{n,m}$ is zero unless $n = m$. Multiplying both sides of Eq. 8.11 by $\sin(m\omega t)$ and integrating over a period yields

$$\frac{2}{T} \int_0^T v(t)\ \sin(m\omega t)\ dt = \sum_n a_n \int_0^T \frac{2}{T} \sin(n\omega t)\ \sin(m\omega t)\ dt$$

Making use of orthogonality relation 8.12, we then obtain

$$a_m = \frac{2}{T} \int_0^T v(t)\ \sin(m\omega t)\ dt$$

These coefficients could be obtained only because we had knowledge of the orthogonality relations for our simple functions.

8.4.1   Orthogonality Relations for Dielectric Waveguides

The orthogonality relations for the unperturbed guided modes can be obtained from the Lorentz reciprocity relation with the perturbation removed. In this case $\epsilon' = \epsilon$ and the right side of Eq. 8.10 equals zero; additionally, both the primed and unprimed fields on the left side of the equation correspond to unperturbed solutions. Let $\mathbf{E}$ and $\mathbf{H}$ correspond to any mode n that can exist on the unperturbed guide and let $\mathbf{E}'$ and $\mathbf{H}'$ correspond to any arbitrary mode m on the same guide. Thus we let

$$\begin{Bmatrix} \mathbf{E} \\ \mathbf{H} \end{Bmatrix} = \begin{Bmatrix} \mathcal{E}_n(x,\ y) \\ \mathcal{H}_n(x,\ y) \end{Bmatrix} e^{-jk_{zn}z}$$

$$\begin{Bmatrix} \mathbf{E}' \\ \mathbf{H}' \end{Bmatrix} = \begin{Bmatrix} \mathcal{E}_m(x,\ y) \\ \mathcal{H}_m(x,\ y) \end{Bmatrix} e^{-jk_{zm}z}$$

Making use of Eq. 8.10 then yields

$$(k_{zn} - k_{zm}) \iint_S (\mathcal{E}_{nt}^* \times \mathcal{H}_{mt} + \mathcal{E}_{mt} \times \mathcal{H}_{nt}^*) \cdot \hat{z}\ dS = 0 \tag{8.13}$$

Note that, provided $k_{zm} \neq k_{zn}$, this implies that the above integral must be zero.

Now the mode m is arbitrary and we are therefore free to replace it with any other mode, say $-m$. We will use the convention that positive integers correspond to forward-traveling waves and negative integers to backward-traveling waves. The mode $-m$ therefore corresponds to a wave traveling in the opposite direction to mode m but having the same magnitude for its propagation constant. That is,

$$k_{z(-m)} \equiv -k_{zm}$$

We have previously shown that the transverse fields of the oppositely traveling waves are related via

$$\mathcal{E}_{-mt} = \mathcal{E}_{mt}$$

$$\mathcal{H}_{-mt} = -\mathcal{H}_{mt}$$

Therefore, substituting $-m$ for m in Eq. 8.10 and making use of the above relations yields

$$(k_{zn} + k_{zm}) \iint_S (-\mathcal{E}_{nt}^* \times \mathcal{H}_{mt} + \mathcal{E}_{mt} \times \mathcal{H}_{nt}^*) \cdot \hat{z}\, dS = 0 \qquad (8.14)$$

which implies that if $k_{zm} \neq k_{zn}$ the above integrand must be zero. Adding Eqs. 8.13 and 8.14 then yields

$$\iint_S (\mathcal{E}_{mt} \times \mathcal{H}_{nt}^*) \cdot \hat{z}\, dS = 0 \qquad n \neq \pm m$$

Let us now examine what happens to the above relation when $n = +m$ or $n = -m$. When $n = +m$ the integrand $\mathcal{E}_{nt} \times \mathcal{H}_{nt}^*$ is simply the complex Poynting power density so that

$$\mathrm{Re}\left[\iint_S (\mathcal{E}_{nt} \times \mathcal{H}_{nt}^*) \cdot \hat{z}\, dS \right] = 2P \, \mathrm{sgn}(n)$$

The quantity $\mathrm{sgn}(n)$ is defined to be $+1$ if n is positive and -1 if n is negative. This term is needed because although $\mathcal{E}_{-nt} = \mathcal{E}_{nt}$, $\mathcal{H}_{-nt} = -\mathcal{H}_{nt}$. Physically, this sign change results because power flow for a backward-traveling wave is in the opposite direction of that for a forward-traveling wave. Further, as was shown in Section 8.2.2, $\mathcal{E}_{nt}$ and $\mathcal{H}_{nt}$ can be chosen to be real so that we need not take the real part of the above expression. Therefore, when $n = +m$ we have

$$\iint_S (\mathcal{E}_{nt} \times \mathcal{H}_{nt}^*) \cdot \hat{z}\, dS = 2P \, \mathrm{sgn}(n) \qquad (8.15)$$

Consider next when $n = -m$. We obtain for the cross-product:

$$\iint_S (\mathcal{E}_{-m} \times \mathcal{H}_m^*) \cdot \hat{z}\, dS$$

Since $\mathcal{E}_{-m} = \mathcal{E}_m$, this result is identical with Eq. 8.15 above.

These results can be generalized compactly for all m and n as

$$\iint_S (\mathcal{E}_m \times \mathcal{H}_n^*) \cdot \hat{z}\, dS = 2P\, \text{sgn}(n)\, \delta_{|m|,|n|} \tag{8.16}$$

Equation 8.16 represents the mode orthogonality relation for dielectric waveguides.

Example 8.1

Compute the normalized fields, $\mathbf{e}_n$ and $\mathbf{h}_n$, for TE modes defined such that $2P = 1$, that is

$$\iint_S (\mathbf{e}_m \times \mathbf{h}_n^*) \cdot \hat{z}\, dS = \text{sgn}(n)$$

Solution

In Section 4.3.5 the relationship between the power P carried by a TE waveguide mode and the amplitude coefficient A multiplying all the fields was shown to be

$$|A|^2 = \frac{4\omega\mu}{W\, k_z\, d_{\text{eff}}}\, P$$

For the fields to be normalized as defined above, we require that $|A|$ be given by

$$|A| = \left(\frac{2\omega\mu}{W\, k_z\, d_{\text{eff}}}\right)^{1/2}$$

8.5 COUPLED EQUATIONS OF MOTION

Let us return to the Lorentz reciprocity relation in the case where the perturbed dielectric constant ϵ' is no longer equal to ϵ.

Let the unperturbed fields $\mathbf{E}$ and $\mathbf{H}$ correspond to the solutions for the nth mode on the unperturbed waveguide. That is,

$$\left.\begin{array}{l} \mathbf{E}(x, y, z) = \mathbf{e}_n(x, y) \\ \mathbf{H}(x, y, z) = \mathbf{h}_n(x, y) \end{array}\right\} e^{-jk_{zn}z} \tag{8.17}$$

We use lower-case symbols to represent these fields indicating that their amplitudes have been normalized in such a manner that the mode carries unity power ($2P = 1$). We wish to represent the perturbed fields $\mathbf{E}'$ and $\mathbf{H}'$ as some superposition of the unperturbed modes. At some fixed arbitrary value of $z = z_0$ the perturbed fields are only a function of x and y. The transverse field components of the unperturbed modes have already been shown to be an orthogonal set in any xy plane so we attempt to express the transverse components of our perturbed fields in terms of these, that is

$$\left.\begin{array}{c} \mathbf{E}'_t(x, y, z_0) \\ \mathbf{H}'_t(x, y, z_0) \end{array}\right\} = \sum_m a_m(z_0) \left\{\begin{array}{c} \mathbf{e}_{mt}(x, y) \\ \mathbf{h}_{mt}(x, y) \end{array}\right\} e^{-jk_{zm}z_0}$$

where the coefficients $a_m(z_0)$ are constants and the summation is over all guided modes. It should be pointed out that, strictly speaking, the above superposition should also include the radiation modes of the type analyzed in Chapter 4. These modes were shown to represent field solutions that are not bound to the waveguide but rather are superpositions of traveling waves moving away from and toward the waveguide core (see Fig. 4.14). Intuitively the reader might suspect that these fields are important in describing geometries in which radiation is either coupled or scattered into or out of the waveguide by some perturbing mechanism. Thus, by neglecting these fields we are implicitly assuming that no such type of coupling occurs.

While the above superposition may be used to represent the perturbed field variation with x and y at $z = z_0'$ at some other different but arbitrary point z along the guide the perturbed field variation with x and y might be different. Therefore, the coefficients a_m must in general vary with z. Thus, for arbitrary z we replace $a_m(z_0)$ above with $a_m(z)$. If we can solve for the coefficients $a_m(z)$ we will have obtained a complete description of the perturbed fields.

We should note that the longitudinal perturbed fields E'_z and H'_z, are obtainable from the transverse components using Maxwell's curl equations. That is,

$$\hat{z}E'_z = \frac{1}{j\omega\epsilon'} (\nabla_t \times \mathbf{H}'_t)$$

$$\hat{z}H'_z = \frac{1}{-j\omega\mu_0} (\nabla_t \times \mathbf{E}'_t)$$

If we substitute the expansions for $\mathbf{E}'_t$ and $\mathbf{H}'_t$ in terms of the unperturbed transverse modes and note that the unperturbed fields satisfy Maxwell's curl equations for the unperturbed geometry defined by ϵ then we obtain

$$\left.\begin{array}{c} E'_z \\ H'_z \end{array}\right\} = \sum_m a_m(z) \left\{\begin{array}{c} \dfrac{\epsilon}{\epsilon'} e_{mz} \\ h_{mz} \end{array}\right\} e^{-jk_{zm}z}$$

For many problems, the perturbed dielectric constant ϵ' is very close to ϵ so that the ratio ϵ/ϵ' can be approximated by unity. Under these conditions, the field expansion simplifies to the following approximate form:

$$\left.\begin{array}{r}\mathbf{E}'(x, y, z) \\ \mathbf{H}'(x, y, z)\end{array}\right\} = \sum_m a_m(z)\left\{\begin{array}{l}\mathbf{e}_m(x, y) \\ \mathbf{h}_m(x, y)\end{array}\right\}e^{-jk_{zm}z} \tag{8.18}$$

Substituting the assumed form for the unperturbed and perturbed fields, Eqs. 8.17 and 8.18, into the Lorentz reciprocity theorem, Eq. 8.10, yields

$$\begin{aligned}\sum_m \left[j\Delta_{n,m}a_m + \frac{da_m}{dz}\right] e^{j\Delta_{n,m}z} &\iint_S (\mathbf{e}_{nt}^* \times \mathbf{h}_{mt} + \mathbf{e}_{mt} \times \mathbf{h}_{nt}^*) \cdot \hat{z}\, dS \\ &= -j\omega \sum_m a_m\, e^{j\Delta_{n,m}z} \iint_S (\epsilon' - \epsilon)\, \mathbf{e}_m \cdot \mathbf{e}_n^*\, dS\end{aligned} \tag{8.19}$$

where

$$\Delta_{n,m} \equiv k_{zn} - k_{zm}$$

and the functional dependence of a_m on z has been omitted for convenience. Applying the orthogonality relation 8.16 to the integral on the left side of the above equation shows that

$$\iint_S (\mathbf{e}_{nt}^* \times \mathbf{h}_{mt} + \mathbf{e}_{mt} \times \mathbf{h}_{nt}^*) \cdot \hat{z}\, dS = 2\, \text{sgn}(n)\, \delta_{m,n}$$

so that the summation over m on the left side of Eq. 8.19 selects out only the value of m equal to n. Thus, we obtain

$$\frac{da_n}{dz} = -j \sum_m a_m e^{j\Delta_{n,m}z}\, C_{m,n} \tag{8.20}$$

where

$$C_{m,n} \equiv \frac{\omega}{2}\, \text{sgn}(n) \iint_S (\epsilon' - \epsilon)\, \mathbf{e}_m \cdot \mathbf{e}_n^*\, dS$$

8.6 COUPLED WAVEGUIDES

The set of coupled differential equations that we have just derived can be used to analyze how energy is transferred from one optical waveguide to another when the two guiding structures are brought into proximity. The geometry of the problem is shown in Figure 8.5. Dielectric waveguides A and B can have arbitrary cross-sectional shapes which are uniform with respect to the propagation direction z.

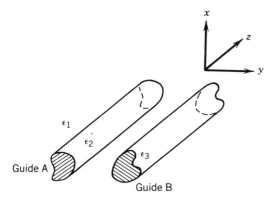

Figure 8.5 Two waveguides brought into proximity.

. To make use of the set of coupled-mode equations 8.20 in analyzing this problem, it is first necessary to define perturbed and unperturbed geometry. Let the unperturbed problem defined by $\epsilon(x, y, z)$ consist of waveguide A with waveguide B removed to infinity as shown in Figure 8.6a. The perturbed problem, defined by $\epsilon'(x, y, z)$, is shown in Figure 8.6b and consists of the geometry of interest, that is, guides A and B in proximity. The perturbation $(\epsilon' - \epsilon)$ then exists only in the region defined by guide B and is given by

$$\Delta\epsilon = (\epsilon' - \epsilon) = (\epsilon_3 - \epsilon_1) \qquad \text{over region defined by guide B}$$

$$= 0 \qquad\qquad\qquad \text{otherwise}$$

This is shown in Figure 8.6c.

Now that the perturbed and unperturbed geometries have been defined, we need merely express the perturbed fields in terms of a superposition of unperturbed fields. Let us further assume that both guides A and B, when isolated from each other, are single-mode structures with field distributions shown diagramatically in Fig. 8.7. We immediately run into a problem in trying to use our coupled-mode formulation. Since the unperturbed problem consists of the single-mode guide A alone, we have only one mode with which to represent our perturbed fields. Clearly, when guides A and B are brought into proximity, a field distribution centered about guide A cannot be a good representation of our perturbed fields since at least part of the energy of the composite waveguide should now be centered about guiding structure B as shown in Figure 8.8.

To determine a better approximation to the perturbed problem, consider what happens to the two waveguides as they are brought into proximity. As the guiding structures are brought into proximity, the fields of each will be perturbed in some manner by the presence of the other dielectric waveguide. Provided that the guides are not too close together, the perturbation will be small. The field distribution on each guide should not be drastically different then when the two guides were far apart. We thus try to represent the perturbed field distribution in the xy plane for the composite guiding

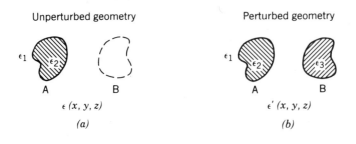

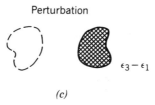

Figure 8.6 (*a*) Unperturbed geometry; (*b*) perturbed geometry; (*c*) perturbation.

system as a superposition of the field distribution existing on each guide in absence of the other. Letting the fields on guides A and B in absence of the other guide be defined by $\mathbf{e}_{A,B}(x, y)$ and $\mathbf{h}_{A,B}(x, y)$ and the respective propagation wavenumbers be $k_{zA,B}$, then we represent the perturbed fields $\mathbf{E}'$ and $\mathbf{H}'$ as

$$\begin{Bmatrix} \mathbf{E}'(x, y, z) \\ \mathbf{H}'(x, y, z) \end{Bmatrix} \cong a_A(z) \begin{Bmatrix} \mathbf{e}_A(x, y) \\ \mathbf{h}_A(x, y) \end{Bmatrix} e^{-jk_{zA}z}$$

$$+ \; a_B(z) \begin{Bmatrix} \mathbf{e}_B(x, y) \\ \mathbf{h}_B(x, y) \end{Bmatrix} e^{-jk_{zB}z}$$

Figure 8.7 Schematic representation of field amplitude variation for the two waveguides when far apart.

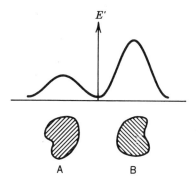

Figure 8.8 Schematic representation of field amplitude variation for the two waveguides when in proximity.

The expansion for the perturbed fields is now in the same form as used in Eq. 8.18 to obtain the coupled set of Eqs. 8.20. We simply replace the summation over the mode numbers $m = 1, 2, \ldots$ with a summation over $m = A, B$. We should note that, strictly speaking, for Eq. 8.20 to be valid the orthogonality relationship

$$\iint_S (\mathbf{e}_{At} \times \mathbf{h}_{Bt}^*) \cdot \hat{z}\, dS = 0$$

must hold. If the waveguides are not too close, then the fields from guide A are small where those from B are appreciable and vice versa. Thus, the above orthogonality requirement is satisfied approximately.

We can, therefore, now make use of the set of coupled relations 8.20. Recall that in this expression the subscript n represents those modes that exist in absence of the perturbation. For our problem, this means the single mode on guide A which exists in absence of guide B. Thus, we have

$$\frac{da_A}{dz} = -j[a_A(z)C_{AA} + a_B(z)e^{j(k_{zA} - k_{zB})z}C_{BA}]$$

where

$$C_{\binom{A}{B}A} = \frac{\omega}{2} \iint_{S_B} (\epsilon_3 - \epsilon_1)\, [\mathbf{e}_{\binom{A}{B}} \cdot \mathbf{e}_A^*]\, dS$$

We note first that the region of integration is over the perturbation which is the region defined by waveguide B. Note that as shown in Figure 8.8, in this region $|\mathbf{e}_A|$ is much smaller than $|\mathbf{e}_B|$ so that the magnitude of the coupling term C_{AA} is much smaller than C_{BA}. Thus,

$$\frac{da_A}{dz} \cong -jC_{BA}a_B(z)e^{j(k_{zA} - k_{zB})z}$$

In the above derivation, we have assumed waveguides A and B to be the unperturbed and perturbing structures, respectively. If we now treat guide B as the unperturbed structure and guide A as the perturbation, we obtain the identical results as above but with A and B everywhere interchanged. That is,

$$\frac{da_B}{dz} = -jC_{AB}a_A(z)e^{-j(k_{zA}-k_{zB})z}$$

where

$$C_{AB} = \frac{\omega}{2} \iint_{S_A} (\epsilon_2 - \epsilon_1)(\mathbf{e}_A \cdot \mathbf{e}_B^*) \, dS$$

and the integration is now over the region defined by guide A.

8.7 FORWARD COUPLING: THE DIRECTIONAL COUPLER

The coupled-mode equations can be written compactly as

$$\frac{da_A}{dz} = -jC_{BA}e^{j\Delta kz}a_B(z) \tag{8.21}$$

$$\frac{da_B}{dz} = -jC_{AB}e^{-j\Delta kz}a_A(z) \tag{8.22}$$

where

$$\Delta k \equiv k_{zA} - k_{zB}$$

Equations 8.21 and 8.22 represent two differential equations in two unknowns. To solve these, let us assume that

$$a_{A,B}(z) = a_{A,B}^0 e^{-j\gamma_{A,B}z}$$

where $a_{A,B}^0$ is an arbitrary constant and $\gamma_{A,B}$ are unknown. Substitution into Eqs. 8.21 and 8.22 converts the differential form to an algebraic one:

$$a_A^0 = \frac{C_{BA}}{\gamma_A}e^{j(\Delta k + \gamma_A - \gamma_B)z}a_B^0 \tag{8.23}$$

$$a_B^0 = \frac{C_{AB}}{\gamma_B}e^{-j(\Delta k + \gamma_A - \gamma_B)z}a_A^0 \tag{8.24}$$

Since Eqs. 8.23 and 8.24 must be true for all z, we require that

$$\gamma_B = \Delta k + \gamma_A \tag{8.25}$$

Substituting for a_B^0 from Eq. 8.24 into Eq. 8.23 and making use of Eq. 8.25 therefore yields

$$\gamma_A(\Delta k + \gamma_A) - C_{BA}C_{AB} = 0$$

or

$$\gamma_A = -\frac{\Delta k}{2} \pm \sqrt{\left(\frac{\Delta k}{2}\right)^2 + C_{BA}C_{AB}}$$

and thus

$$\gamma_B = +\frac{\Delta k}{2} \pm \sqrt{\left(\frac{\Delta k}{2}\right)^2 + C_{BA}C_{AB}}$$

Defining

$$S = \sqrt{\left(\frac{\Delta k}{2}\right)^2 + C_{BA}C_{AB}}$$

then we have for $a_A(z)$

$$a_A(z) = e^{j(\Delta k/2)z}(A_1 e^{jSz} + A_2 e^{-jSz}) \tag{8.26}$$

where A_1 and A_2 are arbitrary constants. To determine their value, we must impose boundary conditions along z appropriate to the problem of interest.

Let us assume that we wish to transfer energy from fiber waveguide B into fiber waveguide A using the integrated optics coupler shown in Figure 8.9. At $z = 0$, all the energy is in mode B, which requires that $a_A(z = 0) = 0$ and from Eq. 8.26 therefore, $A_2 = -A_1$. Thus, $a_A(z)$ is given by

$$a_A(z) = 2jA_1 \, e^{j(\Delta k/2)z} \, \sin(Sz)$$

The coefficient $a_B(z)$ can be obtained from $a_A(z)$ using Eq. 8.21 yielding

$$a_B(z) = \frac{2jA_1}{C_{BA}} e^{-j(\Delta k/2)z} \left[-\frac{\Delta k}{2} \sin(Sz) + jS \cos(Sz) \right]$$

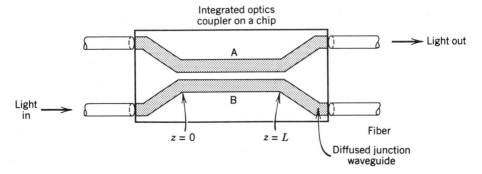

Figure 8.9 Simple integrated-optics coupler.

The power carried in each guide is proportional to the magnitude squared of the coefficient $a_{A,B}(z)$. Letting the power incident on guide B at $z = 0$ be P_0 then

$$P_0 \propto |a_B(0)|^2 = \frac{(2A_1)^2}{|C_{BA}|^2} S^2$$

which specifies the coefficient A_1. Therefore, we have for the power in each waveguide

$$|a_A(z)|^2 = P_0 \frac{|C_{BA}|^2}{S^2} \sin^2(Sz) \equiv P_A(z) \tag{8.27}$$

$$|a_B(z)|^2 = P_0 \left[\left(\frac{\Delta k}{2S} \right)^2 \sin^2(Sz) + \cos^2(Sz) \right] \equiv P_B(z) \tag{8.28}$$

Case I. Phase-Matched Condition ($\Delta k = 0$)

Let us first consider the power in guides A and B under the phase-matched condition $k_{zA} = k_{zB}$, which will automatically be the case if guides A and B are identical. Under this condition, $C_{AB} = C_{BA} \equiv C$ and

$$P_A(z) = P_0 \sin^2(Cz)$$
$$P_B(z) = P_0 \cos^2(Cz) \tag{8.29}$$

Thus, under phase-matched conditions, power transfer occurs periodically with z, as shown in Figure 8.10. The length z_B required for total power transfer under the phase-matched condition is given from Eq. 8.29 by

$$\sin(Cz_B) = 1$$

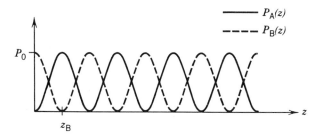

Figure 8.10 Interchange of modal power between the two guides as a function of propagation distance under phase-matched conditions.—, $P_A(z)$; ----, $P_B(z)$.

or

$$z_B = \frac{\pi}{2C} \tag{8.30}$$

Note that as the coupling is reduced, the transfer length increases. This is to be expected from a physical standpoint since the power transferred per unit length from guide B to A is reduced as the coupling is reduced. The sinusoidal beating between the two waveguides is also expected on physical grounds. At the waveguide input, all the guided power is in B. If we assume that the rate of increase in amplitude of the field in guide A at some point z is proportional to the amplitude in B and vice versa, then we expect the power to swap back and forth between the guides. Note that Eqs. 8.21 and 8.22 indicate mathematically precisely the phenomena described above.

Case II. Non-Phase-Matched Condition ($\Delta k \neq 0$)

We note from relations 8.27 and 8.28 that, in general, when $\Delta k \neq 0$, the maximum power transferred to guide A occurs when $Sz = \pi/2$ yielding a value of

$$\frac{(P_A)}{P_0} \text{max} = \frac{|C_{BA}|^2}{C_{AB}C_{BA} + \Delta k^2/4} \tag{8.31}$$

In Figure 8.11 we plot the dependency of the power transferred to guide A as a function of the phase mismatch between the two waveguides, Δk.

When $\Delta k \neq 0$, complete power transfer cannot occur and the relative power in guides A and B as a function of z looks as shown in Fig. 8.12. We observe that if the phase mismatch is increased from zero to a value of $2\sqrt{C_{AB}C_{BA}}$, the maximum power that can be transferred to guide A drops by a factor of 2. Thus, appreciable power transfer requires that

$$|\Delta k| < 2\sqrt{C_{AB}C_{BA}} \tag{8.32}$$

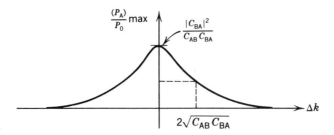

Figure 8.11 Dependence of maximum coupled power on phase mismatch.

Recall that C_{AB} and C_{BA} are the coupling coefficients which measure the overlap of fields A and B. For weak perturbations (guides not too close together), this term is small so that energy is coupled between two waveguides only if the modes for which energy transfer is desired are very close to being phase matched. We see now that the initial assumption that each guide be single mode is not necessary. Let guides B and A each support many modes. If at $z = 0$, the energy is all in mode m_B of guide B this energy can only couple appreciably to a mode in guide A, m_A, for which we have phase matching, that is, $k_{zm_B} = k_{zm_A}$. The amplitude of all other modes is negligible so that we again have the two coupled-mode problem.

One can easily determine at what frequencies and between which modes coupling will occur for any pair of waveguides. Figure 8.13 indicates the dispersion diagram for two fictitious guiding structures, A and B. Since strong coupling occurs only for $k_{zA} = k_{zB}$, we may determine these locations by superimposing the two dispersion diagrams and obtaining the intersection points as shown in Figure 8.14. Thus, in general, if guides A and B are different, strong coupling can occur only at a discrete set of frequencies for which the dispersion curves intersect. Obviously, for identical guides, the dispersion curves for A and B intersect at all frequencies.

The fact that we require phase matching for strong coupling can also be used to create an electrically activated optical switch. Suppose we place an electrode to either

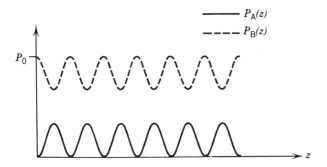

Figure 8.12 Interchange of modal power between the two guides as a function of propagation distance under non-phase-matched conditions. —, $P_A(z)$; ----, $P_B(z)$.

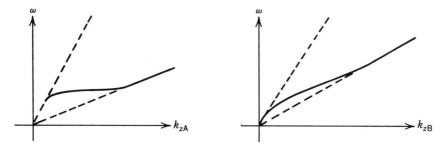

Figure 8.13 Dispersion diagrams for two arbitrary waveguides.

side of guide B as shown in Figure 8.15. If the substrate is electro-optic as is, for example, LiNbO$_3$, then the application of a dc voltage across the electrodes will produce a change in the dielectric constant primarily in the region associated with guide B. With no voltage applied, the two guides are identical so that phase matching exists and if L is chosen properly, the optical energy will be transferred from guide B to A. With an applied voltage, the guides become dissimilar due to the electro-optic effect and energy transfer can be prevented. One can envision an entire switching network as shown in Figure 8.16 for optical fibers based upon the principles described in this section.

Example 8.2

Compute the coupling coefficient C_{AB} for the fundamental TE modes on the two dielectric slab waveguides shown in Figure 8.17.

Solution

The coupling coefficient C_{AB} is

$$C_{AB} = \frac{\omega}{2} \iint_{S_A} (\epsilon_2 - \epsilon_1)(\mathbf{e}_A \cdot \mathbf{e}_B^*) \, dS$$

$$= \frac{\omega(\epsilon_2 - \epsilon_1)W}{2} \int_{-d/2}^{d/2} dx \, \mathbf{e}_A(x) \cdot \mathbf{e}_B^*(x)$$

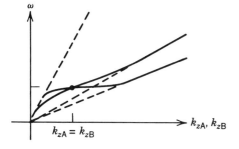

Figure 8.14 Graphical solution for phase-matched operating frequency and wavenumber.

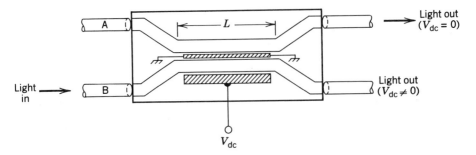

Figure 8.15 Integrated optical fiber electro-optic switch based upon directional coupler analysis.

From Section 4.2.1 and Example 8.1, the normalized electric fields expressed relative to the coordinate system shown are given by

$$\mathbf{e}_A(x) = \hat{y}\,|A|\,\cos(k_{2x}x) \qquad\qquad |x| < d/2$$

$$\mathbf{e}_B(x') = \hat{y}\,|A|\,\cos(k_{2x}d/2)e^{-\alpha_x(x'-d/2)} \qquad x' > d/2$$

or

$$\mathbf{e}_B(x) = \hat{y}\,|A|\,\cos(k_{2x}d/2)e^{-\alpha_x(x+s+d/2)} \qquad x > -(s+d/2)$$

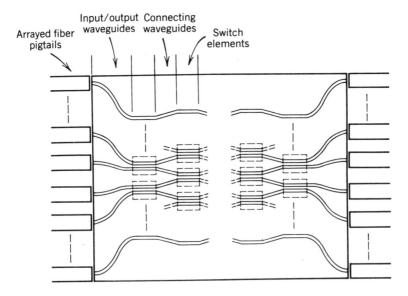

Figure 8.16 Integrated optical fiber switching network. (After Kondo *et al.* (Ref. 8). Copyright 1982, IEEE. Reproduced by permission.)

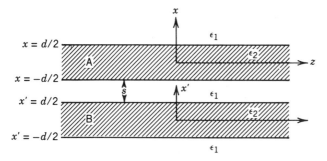

Figure 8.17 Geometry for Example 8.2.

and

$$|A| = \left(\frac{2\omega\mu}{W\, k_z\, d_{\text{eff}}}\right)^{1/2}$$

Substitution of these fields into the above expression for C_{AB} yields

$$C_{AB} = \frac{\omega^2\mu(\epsilon_2 - \epsilon_1)}{k_z\, d_{\text{eff}}} e^{-\alpha_x(s+d/2)} \cos(k_{2x}d/2) \int_{-d/2}^{d/2} dx\, \cos(k_{2x}x)e^{-\alpha_x x}$$

The above integral can be solved by noting that it is of the form

$$\text{Re} \int_{-d/2}^{d/2} dx\, e^{j(k_{2x}+j\alpha_x)x}$$

Performing the above integration and making use of the guidance condition for even TE modes,

$$\tan(k_{2x}d/2) = \alpha_x/k_{2x}$$

then yields

$$C_{AB} = \frac{2\omega^2\mu(\epsilon_2 - \epsilon_1)\,\alpha_x\,\cos^2(k_{2x}d/2)}{k_z\, d_{\text{eff}}\,(\alpha_x^2 + k_{2x}^2)} e^{-\alpha_x s}$$

Observe that coupling increases exponentially with decreasing waveguide separation.

8.8 BACKWARD COUPLING: THE GRATING REFLECTOR

In the last section, we analyzed how a forward-traveling mode in one waveguide could have its power transferred to a forward-traveling mode in a second waveguide. In this section, a general method will be presented for transferring the power in a forward-traveling mode to that of a backward-traveling mode existing within the same guide. This transfer is accomplished by the introduction of a periodic perturbation in waveguide dielectric along the propagation direction z. Figure 8.18a shows the geometry of interest. It consists of a 2-D slab dielectric waveguide having a shallow grating of height h and periodicity Λ. The grating can be fabricated using one of the mask and etch techniques described in Chapter 7. Note that this waveguide may be considered as a perturbation on the simpler uniform slab waveguide shown in Figure 8.18b. The perturbation consists of the square wave-shaped region, $\Delta\epsilon(x, z)$ shown in Figure 8.18c. When either h or $\epsilon_2 - \epsilon_1$ or both are small quantities, then the perturbed problem should be described accurately using our coupled-mode formalism. The perturbation can be described mathematically as

$$\Delta\epsilon(x, z) = u(x)\, \Delta\epsilon(z)$$

where

$$u(x) = 1 \qquad |x - d/2| < h/2$$
$$= 0 \qquad \text{otherwise}$$

and

$$\Delta\epsilon(z) = \Delta\epsilon_1 \cos\frac{2\pi}{\Lambda} z + \Delta\epsilon_2 \cos\frac{4\pi}{\Lambda} z + \cdots + \Delta\epsilon_n \cos\frac{2n\pi}{\Lambda} z \qquad (8.33)$$

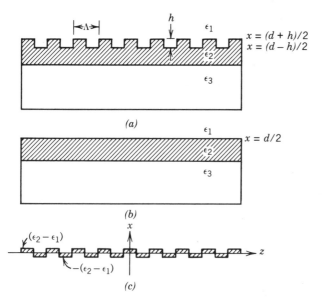

Figure 8.18 (a) The grating reflector; (b) unperturbed problem; (c) perturbation.

where for a rectangular pulse

$$\Delta\epsilon_n = \frac{4(\epsilon_2 - \epsilon_1)}{\pi n}(-1)^{(n-1)/2} \qquad n \text{ odd}$$

$$= 0 \qquad\qquad\qquad n \text{ even}$$

Using superposition, the influence of each harmonic of the perturbation may be analyzed individually and the final result obtained by summing the effects of all terms. We therefore will consider explicitly only the influence of the first harmonic.

Let us assume that the unperturbed guide of Figure 8.18b is single mode, capable of supporting one forward mode, say n, and one backward mode $-n$ where n is a positive integer. Thus from Eq. 8.18, the perturbed fields are assumed to be of the form

$$\begin{Bmatrix} \mathbf{E}'(x, y, z) \\ \mathbf{H}'(x, y, z) \end{Bmatrix} = a_n(z) \begin{Bmatrix} \mathbf{e}_n(x, y) \\ \mathbf{h}_n(x, y) \end{Bmatrix} e^{-jk_{zn}z}$$

$$+ a_{-n}(z) \begin{Bmatrix} \mathbf{e}_{-n}(x, y) \\ \mathbf{h}_{-n}(x, y) \end{Bmatrix} e^{-jk_{z(-n)}z} \tag{8.34}$$

The set of coupled-mode equations given by Eq. 8.20 therefore reduce to a sum on $m = \pm n$. Thus, we have

$$\frac{da_n}{dz} = -j(a_n C_{n,n} + a_{-n} e^{j\Delta_{n,-n}z} C_{-n,n}) \tag{8.35}$$

and

$$\frac{da_{-n}}{dz} = -j(a_{-n} C_{-n,-n} + a_n e^{j\Delta_{-n,n}z} C_{n,-n})$$

where

$$C_{nn} = \frac{\omega}{2}\Delta\epsilon_1 \cos(2\pi z/\Lambda)\int_0^W dy \int_{(d-h)/2}^{(d+h)/2} dx |\mathbf{e}_n(x, y)|^2 \tag{8.36}$$

with

$$C_{-n,n} = C_{n,n}$$

$$C_{n,-n} = C_{-n,-n} = -C_{n,n}$$

and

$$\Delta_{n,-n} = k_{zn} - k_{z(-n)} = 2k_{zn}$$

$$\Delta_{-n,n} = -2k_{zn}$$

If the grating height h is small, then the fields $\mathbf{e}_n(x, y)$ and $\mathbf{e}_n^*(x, y)$ may be assumed to be constant over the integration with respect to x in Eq. 8.36 and equal to their value at $x = d/2$. Thus,

$$C_{n,n} \cong C_0 \cos\left(\frac{2\pi}{\Lambda} z\right) \tag{8.37}$$

where

$$C_0 = \frac{\omega}{2} \Delta\epsilon_1 h \int_0^w dy \left| \mathbf{e}_n\left(\frac{d}{2}, y\right) \right|^2$$

The pair of coupled equations 8.35 can then be written compactly as

$$\frac{da_n}{dz} = -jC_0(a_n \cos \gamma z + a_{-n}e^{j2k_{zn}z} \cos \gamma z) \tag{8.38}$$

$$\frac{da_{-n}}{dz} = +jC_0(a_{-n} \cos \gamma z + a_n e^{-j2k_{zn}z} \cos \gamma z) \tag{8.39}$$

where

$$\gamma \equiv \frac{2\pi}{\Lambda}$$

To solve for a_n and a_{-n} we first make several observations concerning the behavior of these two coefficients. Note that as the periodic perturbation approaches zero, then, from Eq. 8.37, C_0 approaches zero. In this limit, $da_n/dz = da_{-n}/dz = 0$, implying that a_n and a_{-n} are constants. For a small but finite perturbation, a_n and a_{-n} are no longer constants but are slowly varying functions of z. This is of key importance in obtaining a solution to coupled Eqs. 8.38 and 8.39. Suppose we attempt to solve the first of these two equations by integrating both sides with respect to z. Note that the first term on the right side of Eq. 8.38 consists of a slowly varying function, $a_n(z)$, multiplied by a rapidly oscillating one, $\cos \gamma z$, as shown in Figure 8.19. Regardless of the exact functional form of $a_n(z)$, the oscillating nature of the term $\cos \gamma z$ causes the product of these two terms to average to nearly zero when integrated over z. Consider, however, the second term on the right side of Eq. 8.38 which can be rewritten as

$$a_{-n}(z) e^{j2k_{zn}z} \cos \gamma z = a_{-n}(z)\left[\frac{e^{j(\gamma + 2k_{zn})} + e^{-j(\gamma - 2k_{zn})z}}{2}\right]$$

Both the real and imaginary parts of the first term in square brackets are rapidly oscillating and therefore their product with $a_{-n}(z)$ also tends to give negligible contribution when integrated over z. Care must be taken with the second term in square

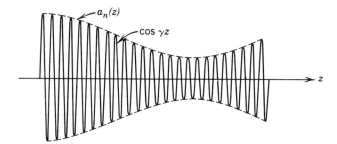

Figure 8.19 The product of a slowly and rapidly varying function yields a quantity that tends to average to zero.

brackets, however. If the grating periodicity is chosen such that

$$2k_{zn} - \gamma \equiv \delta k$$

where δk is a small quantity, then this term can also be slowly varying and thus the integral of its product with $a_{-n}(z)$ will not be negligible. We are thus led to the conclusion that when the grating periodicity is chosen so that δk is small, then Eq. 8.38 can be written approximately as

$$\frac{da_n}{dz} = -j\frac{C_0}{2} a_{-n}e^{j\delta kz} \tag{8.40}$$

By precisely the same arguments, Eq. 8.39 becomes

$$\frac{da_{-n}}{dz} = +j\frac{C_0}{2} a_n e^{-j\delta kz} \tag{8.41}$$

Note that Eqs. 8.40 and 8.41 are of a form that is analogous to the pair of coupled equations 8.21 and 8.22 of the last section. Thus we again assume solutions of the form

$$a_{\pm n}(z) = a^0_{\pm n}e^{j\gamma^+ z}$$

from which we obtain in analogous fashion

$$\gamma^+ = \frac{\delta k}{2} \pm jQ$$

$$\gamma^- = -\frac{\delta k}{2} \pm jQ$$

where

$$Q \equiv \sqrt{(C_0/2)^2 - (\delta k/2)^2} \tag{8.42}$$

Thus, $a_{-n}(z)$ is of the form

$$a_{-n}(z) = e^{-j(\delta kz/2)} (A_1 e^{Qz} + A_2 e^{-Qz})$$

To obtain the unknown constants A_1 and A_2 we must impose boundary conditions. With reference to Figure 8.20, let us therefore assume that the forward-traveling wave is incident from a uniform section of waveguide that exists for $z < 0$ upon a section of grating waveguide of length L. The periodic variation in permittivity is anticipated to continually reflect a small portion of the forward-propagating wave as it travels to the right through the grating region thereby coupling the power of this mode into a backward-propagating wave. Beyond $z = L$, however, no means for backward coupling exists and therefore, at the right-hand edge of the grating waveguide, $z = L$, only the forward-propagating mode exists. Thus, we have

$$a_{-n}(L) = 0$$

which implies that

$$A_2 = -A_1 e^{2QL}$$

and therefore the amplitude of the backward-traveling wave is given by

$$a_{-n}(z) = A_1' e^{-j(\delta kz/2)} \sinh[Q(z - L)] \tag{8.43}$$

The corresponding power in the backward-traveling wave is proportional to $|a_{-n}(z)|^2$, which from above can be written as

$$P_-(z) \equiv |a_{-n}(z)|^2 = |A_1'|^2 \sinh^2[Q(z - L)] \tag{8.44}$$

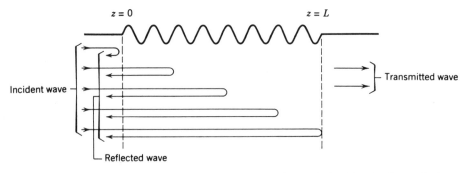

Figure 8.20 Backward scattering of the incident wave by a periodic perturbation. Note that for $z > L$, there is no reflected wave since no scattering source exists to generate it.

The amplitude of the forward-traveling wave, $a_n(z)$, is obtained from $a_{-n}(z)$ using Eq. 8.41. We find straightforwardly that the corresponding power in the wave $P_+(z)$ is given by

$$P_+(z) \equiv |a_n(z)|^2$$

$$= \frac{4|A_1'|^2}{C_0^2} \left\{ Q^2 \cosh^2[Q(z - L)] + \left(\frac{\delta k}{2}\right)^2 \sinh^2[Q(z - L)] \right\} \tag{8.45}$$

To evaluate the constant A_1' we must specify the power incident upon the grating region from the left. This is the power contained in the forward-traveling wave at $z = 0$. Therefore, defining this power as P_{inc}, we require

$$P_+(0) = P_{inc}$$

which from Eq. 8.45 yields

$$|A_1'|^2 = \frac{P_{inc} C_0^2}{4 \left[Q^2 \cosh^2 QL + \left(\frac{\delta k}{2}\right)^2 \sinh^2 QL \right]} \tag{8.46}$$

Case I. Phased-Matched Condition

Let us first consider the situation when the grating periodicity Λ is chosen such that $\delta k = 0$. Then $Q = C_0/2$ and

$$P_+(z) = P_{inc} \cosh^2[C_0(z - L)/2]/\cosh^2(C_0 L/2)$$

$$P_-(z) = P_{inc} \sinh^2[C_0(z - L)/2]/\cosh^2(C_0 L/2)$$

Note that since $\cosh^2 x - \sinh^2 x = 1$ we have

$$P_+(z) - P_-(z) = P_{inc}/\cosh^2(C_0 L/2) = \text{const}$$

or equivalently

$$\frac{d}{dz}[P_+(z) - P_-(z)] = 0 \tag{8.47}$$

Equation 8.47 is a statement of conservation of power which can be interpreted with the aid of Figure 8.21. Consider the two planes located along the waveguide at points z and $z + \Delta z$. Also shown is the power in the forward- and backward-traveling waves at these two planes. As the forward-propagating mode travels between z and $z + \Delta z$,

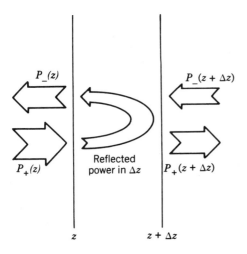

Figure 8.21 Power balance over a differential length of grating reflector.

it gives up a portion of its power to the backward-traveling mode. Thus

$$\underbrace{P_+(z) - P_+(z + \Delta z)}_{\substack{\text{Power given up by the} \\ \text{forward-propagating mode} \\ \text{in going from } z \text{ to } z + \Delta z \\ \text{due to reflections}}} = \underbrace{P_-(z) - P_-(z + \Delta z)}_{\substack{\text{Power gained by the} \\ \text{backward-propagating mode} \\ \text{between } z \text{ and } z + \Delta z \text{ due to} \\ \text{reflections}}} \quad (8.48)$$

Dividing both sides of Eq. 8.48 by Δz and taking the limit as Δz approaches zero yields the results given in Eq. 8.47.

The variation in power with z for both waves is shown in Figure 8.22. The reflectivity

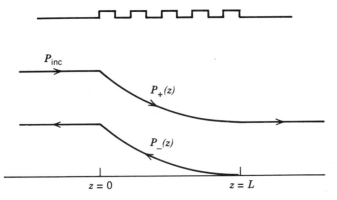

Figure 8.22 Variation of incident and reflected power with distance.

of the grating under phase matched conditions as viewed at the input is given by

$$r_{\text{grating}} = \frac{P_-(0)}{P_+(0)} = \tanh^2(C_0 L/2)$$

Note that as the grating length approaches infinity the tanh function and thus the reflectivity approaches unity.

Case II. Non-Phase-Matched Condition

Let us examine the sensitivity of the grating reflectivity to the phase-match parameter δk. From Eqs. 8.44 and 8.45, when $\delta k \neq 0$

$$r_{\text{grating}} = \frac{(C_0/2)^2 \sinh^2 QL}{Q^2 \cosh^2 QL + (\delta k/2)^2 \sinh^2 QL} \tag{8.49}$$

with

$$Q = \sqrt{(C_0/2)^2 - (\delta k/2)^2}$$

A plot of r_{grating} versus δk would show that this quantity decreases smoothly as δk is increased from zero to C_0 but does not go to zero in this range. To find the first null in r_{grating} we must in fact increase δk beyond the value of C_0. For $\delta k > C_0$ the quantity Q becomes jQ' where

$$Q' = \sqrt{(\delta k/2)^2 - (C_0/2)^2} \tag{8.50}$$

Further, noting that $\sinh(jx) = j \sin x$, then the numerator of Eq. 8.49 is proportional to $\sin^2(Q'L)$. This quantity goes to zero for

$$Q'L = p\pi \qquad p = 1, 2, \ldots$$

From Eq. 8.50, the first null, for which $p = 1$, corresponds to

$$\delta k = \frac{2\pi}{L} \sqrt{1 + (C_0 L/2\pi)^2} \tag{8.51}$$

A plot of r_{grating} versus δk over an extended range is shown in Figure 8.23.

Example 8.3

Compute the coupling coefficient C_0 for the fundamental TE mode on a symmetric slab waveguide having a grating height $h \ll d$.

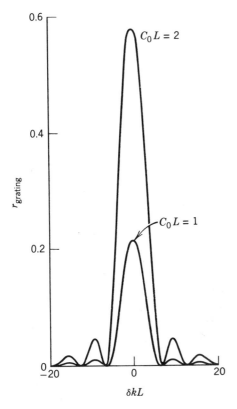

Figure 8.23 Grating reflectivity as a function of normalized phase mismatch for two values of normalized coupling coefficient.

Solution

For the slab waveguide, the fields are independent of y. Therefore,

$$C_0 \cong \frac{\omega \, \Delta\epsilon}{2} \, hW \, |\mathbf{e}(x = d/2)|^2$$

The normalized field $\mathbf{e}(x)$ is given in Example 8.2 which evaluated at $x = d/2$ yields

$$\mathbf{e}(x = d/2) = \hat{y} \left(\frac{2\omega\mu}{Wk_z d_{\text{eff}}} \right)^{1/2} \cos(k_{2x} d/2)$$

Substituting $\mathbf{e}(x = d/2)$ into the expression for C_0 above, we obtain

$$C_0 = \frac{\omega^2 \mu \, \Delta\epsilon}{k_z d_{\text{eff}}} \cos^2(k_{2x} d/2) h$$

Example 8.4   The Waveguide Grating as a Filter

Suppose the optical frequency of the light source exciting our guided mode is varied. What will the influence be upon the grating reflectivity? We consider first the situation when the excitation frequency is chosen so that the phase-matched condition is satisfied. That is,

$$\delta k = 2k_z(\omega_0) - \gamma = 0$$

The proper frequency ω_0 may be ascertained by the graphical construction shown in Figure 8.24. The waveguide dispersion diagram $k_z(\omega)$ is first plotted. Next, the plot is reversed and shifted by γ, yielding the curve $-k_z(\omega) + \gamma$, and superimposed upon the first plot. The intersection points satisfy the relation

$$k_z(\omega_0) = -k_z(\omega_0) + \gamma \tag{8.52}$$

which is equivalent to $\delta k = 0$ and defines the phase-matched frequency ω_0.

Suppose the source frequency is now changed by a small amount, $\Delta\omega$. The corresponding change in k_z will be

$$dk_z = \left.\frac{\partial k_z}{\partial \omega}\right|_{\omega_0} \Delta\omega = \frac{\Delta\omega}{v_g}$$

In practice, it is usually the source free-space wavelength λ that is specified. Frequency is related to wavelength by

$$\lambda = c/f$$

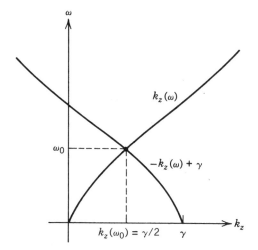

Figure 8.24 Graphical construction to determine phase-matched operating frequency for a grating reflector.

Thus, a change in frequency of Δf about f_0 corresponds to a change in wavelength $\Delta\lambda$ given by

$$\Delta\lambda = -\frac{c}{f_0^2}\Delta f$$

so that

$$|dk_z| = \frac{2\pi|\Delta f|}{v_g} = \frac{2\pi f_0^2 \Delta\lambda}{cv_g} = \frac{\omega_0}{v_g}\frac{\Delta\lambda}{\lambda_0} \tag{8.53}$$

where λ_0 is the phase-matched free-space wavelength. Frequency can be eliminated from the above expression by noting that

$$\frac{\omega_0}{k_{zn}} = v_p$$

and for δk small,

$$k_{zn} \cong \gamma/2$$

Therefore,

$$|dk_z| \cong \frac{v_p}{v_g}\frac{\gamma}{2}\frac{\Delta\lambda}{\lambda_0} = \frac{\pi v_p}{\Lambda v_g}\frac{\Delta\lambda}{\lambda_0}$$

The corresponding phase mismatch δk is therefore

$$|\delta k| = |2k_{zn} - \gamma| = |2[k_{zn}(\omega_0) + dk_z] - \gamma| = 2|dk_z|$$

or

$$|\delta k| = \frac{2\pi}{\Lambda}\frac{v_p}{v_g}\frac{\Delta\lambda}{\lambda_0}$$

To ascertain the fractional change in wavelength required for the reflectivity to go to zero, Eq. 8.51 is used yielding

$$\frac{\Delta\lambda}{\lambda_0} = \frac{v_g}{v_p}\frac{\Lambda}{L}\sqrt{1 + (C_0 L/2\pi)^2} \tag{8.54}$$

Since the maximum reflectivity is given by

$$r_{max} = \tanh^2(C_0 L/2)$$

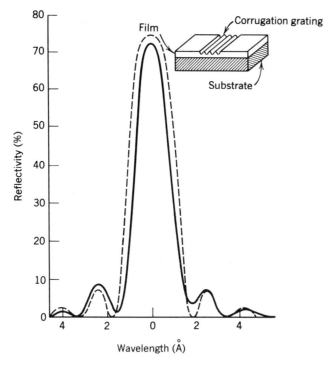

Figure 8.25 Comparison of theoretical and experimental results for a waveguide grating ($L = 1000 \Lambda$). (After Flanders *et al.* (Ref. 6). Copyright 1974, Bell Laboratories, Inc. Reprinted by permission of the authors and the American Institute of Physics.)

Eq. 8.54 can be rewritten as

$$\frac{\Delta\lambda}{\lambda_0} = \frac{v_g}{v_p}\frac{\Lambda}{L}\left\{1 + \frac{1}{\pi^2}[\tanh^{-1}(r_{max}^{1/2})]^2\right\}^{1/2} \tag{8.55}$$

Equation 8.55 states that the bandwidth of the filter is inversely proportional to the grating length but for fixed length increases as the reflectivity is increased. A comparison between experiment and the above theory is presented in Figure 8.25 for a 1000Λ grating filter. Note the filter bandwidth is only 4 Å.

PROBLEMS

8.1 Verify that the relations given in Eq. 8.7 hold for the even TE modes on a symmetric slab waveguide.

8.2 Verify, as discussed in Section 8.2.2, that for odd TE modes on the slab dielectric waveguide the transverse fields can be chosen purely real and the longitudinal fields purely imaginary.

8.3   Verify orthogonality relation 8.16 when $m \neq n$ for two arbitrary even TE modes on a symmetric slab waveguide.

8.4   Two slab dielectric waveguides A and B of thickness d each with core permittivity ϵ_2 are placed a distance s apart. Compute the coupling coefficient C_{AB} between the first odd TE mode on guide A and on guide B.

8.5   In deriving Eq. 8.19 it was assumed that the perturbed permittivity ϵ' was sufficiently close to the unperturbed permittivity ϵ that the ratio ϵ/ϵ' could be approximated by unity. Show that when this is not a valid assumption the term $\mathbf{e}_m \cdot \mathbf{e}_n^*$ in Eq. 8.19 must be replaced by the following:

$$(\mathbf{e}_{mt} + \hat{z}\, \epsilon/\epsilon'\, e_{mz}) \cdot \mathbf{e}_n^*$$

8.6   Show that the coupling coefficient for the fundamental TE modes given in Example 8.2 can be simplified for the symmetric guide to

$$C_{AB} = \frac{\alpha_x k_{2x}^2\, e^{-\alpha_x s}}{k_z(d/2)(1 + 2/\alpha_x d)(\alpha_x^2 + k_{2x}^2)}$$

Hint. Make use of the guidance condition and dispersion relations in regions 1 and 2.

8.7   A directional coupler is to be fabricated using two identical slab dielectric waveguides having the following specifications:

1. Waveguide core thickness: $d = 0.5$ μm
2. Waveguide center-to-center spacing: $s + d = 1.5$ μm
3. Core refractive index: $n_2 = 1.4$
4. Surrounding refractive index: $n_1 = 1.0$
5. Operating free-space wavelength: $\lambda_0 = 1.2$ μm

Compute the value of the coupling coefficient C_{AB} for the fundamental TE modes.

8.8   For the directional coupler of problem 8.8, what must the waveguide separation s be for complete power transfer to occur over a distance of 1 mm?

8.9   Show that the grating coupling coefficient C_0 for the fundamental symmetric TE mode can be simplified on a sinusoidal grating of height h to yield

$$C_0 = \frac{k_{2x}^2\, h}{k_z\, d_{\text{eff}}}$$

Use the hint given in problem 8.6.

8.10   A symmetric slab waveguide has the following specifications:

1. $n_1 = 1.4$, $n_2 = 1.434$, $n_3 = 1.4$
2. $\lambda_0 = 1$ μm
3. $d = 1.02$ μm

If the guiding region has a sinusoidal grating of height $h = 0.1$ μm, how long must the grating region be for 90% reflectivity of the fundamental TE mode?

8.11 Using Eqs. 8.41 and 8.43, derive Eq. 8.45.

8.12 A slab dielectric waveguide grating filter is to have the following properties for the fundamental TE mode:

1. $\lambda_0 = 1$ μm
2. Null to null bandwidth, $2\Delta\lambda = 10$ Å
3. Guide thickness $d = 3$ μm
4. $n_1 = 1.0$, $n_2 = 1.561$, $n_3 = 1.4$
5. Maximum reflectivity $r_{max} = 95\%$

(a) What must be the grating periodicity Λ for maximum reflectivity at the given operating wavelength?

(b) What must be the number of grooves N to meet the above specifications? You may assume that $v_p \cong v_g$ (Why?).

REFERENCES

1. Adler, R. B., "Waves on inhomogenous cylindrical structures." *Proceedings of the IRE* 40 (1952): 339–348.

2. Arnaud, J. A. "Transverse coupling in fiber optics, Part I. Coupling between trapped modes." *Bell System Technical Journal* 53 (1974): 217–224.

3. Barnoski, M. K. *Introduction to Integrated Optics*. New York: Plenum, 1973.

4. Bresler, A. D., Joshi, G. H., and Marcuvitz, N. "Orthogonality properties for modes in passive and active uniform wave guides." *Journal of Applied Physics* 29 (1958): 794–799.

5. Elachi, C., and Yeh, C. "Frequency selective coupler for integrated optics systems." *Optics Communications* 7 (1973): 201–204.

6. Flanders, D. C., *et al.* "Grating filters for thin-film optical waveguides." *Applied Physics Letters* 24 (1974): 195–196.

7. Harrington, R. F. *Time-Harmonic Electromagnetic Fields*. New York: McGraw-Hill, 1961.

8. Kondo, M., *et al.* "Integrated optical switch matrix for single-mode fiber networks." *IEEE Transactions on Microwave Theory and Techniques* MTT-30 (1982): 1747–1753.

9. Marcuse, D. *Light Transmission Optics*. New York. Van Nostrand-Reinhold, 1972.

10. Marcuse, D. "The coupling of degenerate modes in two parallel dielectric waveguides." *Bell System Technical Journal* 50 (1971): 1791–1816.

11. Snyder, A. W. "Coupled-mode theory for optical fibers." *Journal of the Optical Society America* 62 (1972): 1267–1277.

12. Tamir, T., ed. *Integrated Optics,* New York: Springer-Verlag, 1975.

13. Taylor, H. F., and Yariv, A. "Guided wave optics." *Proceedings of the IEEE* 62 (1974): 1044–1059.

14. Yariv, A. "Coupled-mode theory for guided-wave optics." *IEEE Journal of Quantum Electronics* QE-9 (1973): 919–933.

CHAPTER 9

Bragg Scattering

9.1 THE BRAGG CONDITION

Thus far our investigation of guided-wave phenomena has been limited to those geometries in which propagation occurs along a single direction, defined as z. For a number of important integrated optics devices, it is necessary to change the propagation direction of the guided wave in response to the application of an audio- or microwave-frequency electrical signal. The interaction of the electrical and optical signals occurs either through a direct perturbation of the optical properties of the dielectric medium by the applied electric field (electro-optic effect) or through conversion of the electrical signal into an acoustical disturbance which subsequently modifies the optical properties (acousto-optic effect). To obtain some physical insight into the interaction mechanism, we first consider the somewhat simpler but analogous problem of the bulk-wave acousto-optic modulator.

With reference to Figure 9.1, the bulk-wave modulator consists of an acousto-optic material having a piezoelectric transducer bonded to one end. The piezoelectric transducer converts a time-harmonic rf electrical signal applied to its terminals into a time-harmonic stress applied at the end of the sample. This stress results in the launching of an acoustic wave which propagates along the y axis. For clarity, let us assume that the piezoelectric transducer launches a compressional or so-called longitudinal acoustic wave along the y axis, much in the same fashion as a loudspeaker creates a sound wave. For any time t_0 a "snapshot" of the crystal therefore reveals a periodic fluctuation in the interatomic spacing and a corresponding small variation in the refractive index. For most materials, the acousto-optic effect is small and therefore the perturbation of the refractive index from its value in absence of a signal is small.

Further, because the acoustic-wave velocity is typically about 5 orders of magnitude less than that of the optical beam passing through it, the position of the induced periodic phase grating remains essentially unchanged over the transit time required for light to pass through the grating. Therefore, for many applications, the generated grating may be treated as stationary.

To analyze the effect of the generated grating let us examine the consequences of a plane wave incident upon a stack of dielectric layers at an angle θ_i as shown in Figure 9.2. Note that to be consistent with the usual notation for this type of problem, θ_i has been redefined as the angle between the incident wave and the planes of the dielectric stack.

Such a stack can represent a true series of dielectric layers or an idealization of a periodic refractive-index variation induced by electro-optic or acousto-optic effects

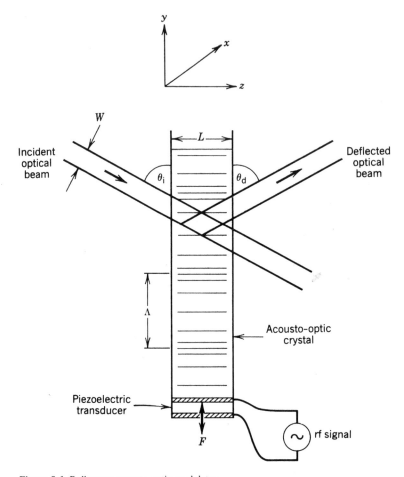

Figure 9.1 Bulk-wave acousto-optic modulator.

such as those discussed previously. We further assume that the fractional difference in dielectric constant, $(\epsilon_2 - \epsilon_1)/\epsilon_2$, is small. Under these circumstances ray theory can readily be applied to determine the angle of incidence required for strong reflection. Since ϵ_2 is close to ϵ_1, we can neglect any change in propagation angle as the incident wave passes from a region of permittivity ϵ_1 to ϵ_2. At each boundary, however, a small portion of the incident power will be reflected due to the discontinuity in dielectric. We are not concerned with the actual value of this reflection coefficient other than to require that it is small. With respect to Figure 9.3, we have already shown from the analysis of the prism coupler that the relative phase difference between successive rays contributing to the pencil beam A is given by

$$|\Delta\phi| = 2k_y\Lambda$$

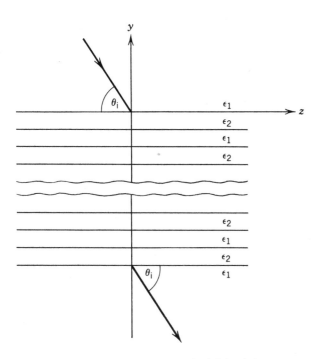

Figure 9.2 Plane wave incident on a stack of dielectric layers.

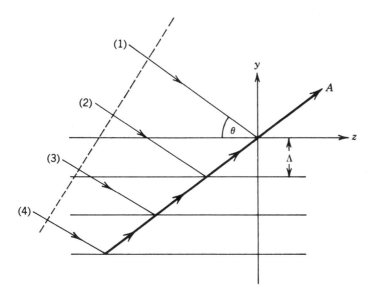

Figure 9.3 Contribution of several incident rays to deflected output ray A.

For all of these contributions to add in phase requires, therefore, that

$$2k_y\Lambda = 2n\pi$$

or with respect to the angle θ shown in Figure 9.3,

$$2k \sin \theta = \frac{2n\pi}{\Lambda} \equiv n\gamma \qquad (9.1)$$

The quantity γ represents the magnitude of an equivalent wavevector associated with the periodic perturbation. When the periodicity Λ is greater than or equal to one-half of the optical wavelength, then a least-one solution to Eq. 9.1 exists. The solution for $n = 1$ is called the first-order Bragg condition and is given by

$$2k \sin \theta_B = \gamma \qquad (9.2)$$

where θ_B is defined as the Bragg angle.

A graphical interpretation of Eq. 9.2 is given in Figure 9.4. Note that the incident and deflected **k** vectors are required to lie on a circle of "radius" $k = \omega\sqrt{\mu_0\epsilon}$. For fixed optical and acoustic frequencies, both k and γ are fixed; hence the Bragg angle is uniquely defined. Based upon the above analysis, if $\theta \neq \theta_B$ strong deflection will not occur.

Example 9.1

An acoustic wave having a velocity $v_{ac} = 3000$ m/s is launched in an acousto-optic crystal having refractive index $n = 2.25$ by means of a bulk-wave transducer excited by an rf of 20 MHz. Compute the Bragg angle θ_B for an indicent He–Ne laser beam.

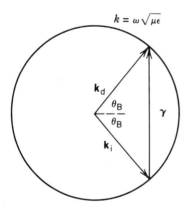

Figure 9.4 Graphical interpretation of the Bragg condition.

Solution

The acoustic wavelength Λ is related to the applied rf signal by

$$f_{rf} \Lambda = v_{ac}$$

Thus,

$$\Lambda = 3 \times 10^3/20 \times 10^6 = 150 \ \mu m$$

In comparison the optical wavelength in the crystal, λ, is

$$\lambda = \lambda_0/n = 0.63 \times 10^{-6}/2.25 = 0.28 \ \mu m$$

The Bragg condition, Eq. 9.2, can be written as

$$\sin \theta_B = \frac{\lambda}{2\Lambda}$$

which gives

$$\theta_B \cong 0.05°$$

Note that the Bragg angle is quite small because the acoustic wavelength is much greater than the optical wavelength.

9.2 THE BULK-WAVE ACOUSTO-OPTIC BEAM DEFLECTOR

The above analysis of Bragg deflection is based upon the assumption that the acoustic wave is an ideal plane wave, that is, its spatial extent transverse to the acoustic propagation direction is infinite. In reality, the beam must have some finite width L. Based upon our analysis of plane-wave diffraction in Section 2.9, we have shown that such a finite aperture beam can be represented by a superposition of true plane waves having an angular spread, $\delta\phi = 2\Lambda/L$. Recall that $\delta\phi$ represented the angular spread between the first set of nulls in the amplitude of the Fourier spectrum. A more conservative measure of the angular spread, over which the amplitude of the plane-wave spectrum is roughly constant, is taken in the literature to be one-half the above value or

$$\delta\phi_{max} \cong \Lambda/L \tag{9.3}$$

The distribution in amplitude of these plane waves is shown schematically in Figure 9.5, which has been normalized to unity for convenience.

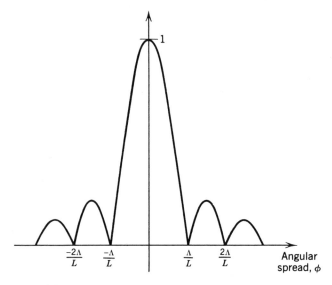

Figure 9.5 Plane-wave spectrum of an acoustic beam with a uniform aperture of width L.

As a consequence of this spread in acoustic propagation directions, it now becomes possible for Bragg deflection to occur over a range of frequencies. This result is shown graphically in Figure 9.6a where the Bragg condition is satisfied at two arbitrary acoustic frequencies f_1 and f_2. At acoustic frequency f_1, the incident optical $\mathbf{k}$ vector $\mathbf{k}_i$, is deflected into the direction $\mathbf{k}_{d1}$ by the acoustic-beam plane-wave component $\boldsymbol{\gamma}_1$. Similarly, at f_2, $\mathbf{k}_i$ is deflected to $\mathbf{k}_{d2}$ by $\boldsymbol{\gamma}_2$. Note that from Figure 9.5, provided that the angular spread $\delta\phi$ between $\boldsymbol{\gamma}_1$ and $\boldsymbol{\gamma}_2$ is much less than Λ/L, the amplitude of the acoustic plane waves associated with these two components are comparable and thus

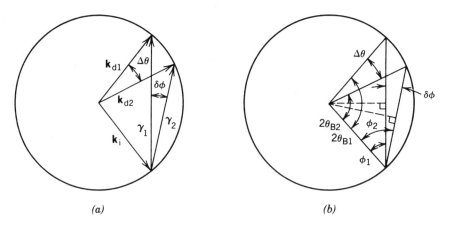

Figure 9.6 (a) Satisfying the Bragg condition at two frequencies, f_1 and f_2; (b) deflection angles, θ_{B1} and θ_{B2}.

the optical waves deflected along $\mathbf{k}_{d1}$ and $\mathbf{k}_{d2}$ should have nominally equal amplitudes.

Using Figure 9.6*b*, the angle of deflection at each corresponding frequency f_i is twice the Bragg angle, which from Eq. 9.2 is given by

$$2\theta_{Bi} = 2 \sin^{-1}\left(\frac{2\pi}{2k\Lambda_i}\right) \cong \frac{\lambda}{\Lambda_i} = \frac{f_i \lambda}{v_{ac}}$$

The difference in optical deflection angle $\Delta\theta$ between two acoustic frequencies separated by Δf is therefore

$$\Delta\theta = \frac{\Delta f \lambda}{v_{ac}} \tag{9.4}$$

Example 9.2

Compute the maximum bandwidth for the acousto-optic deflector shown in Figure 9.1.

Solution

The maximum allowed angular spread $\delta\phi$ for an acoustic beam of width L is equal to from Eq. 9.3

$$\delta\phi_{max} = \frac{\Lambda}{L}$$

If the corresponding change in optical beam direction $\Delta\theta$ can be found, then Eq. 9.4 can be used to obtain the corresponding frequency spread of bandwidth Δf. Now from Figure 9.6 we see that

$$\theta_{B1} + \phi_1 + \frac{\pi}{2} = \pi$$

and

$$\theta_{B2} + \phi_2 + \frac{\pi}{2} = \pi$$

thus

$$2(\theta_{B1} - \theta_{B2}) \equiv \Delta\theta = 2\delta\phi$$

The bandwidth Δf is thus given by

$$\frac{\Delta f \lambda}{v_{ac}} = 2\delta\phi_{max} = \frac{2\Lambda}{L} \tag{9.5}$$

and therefore

$$\Delta f = \frac{2\Lambda v_{ac}}{L\lambda} \tag{9.6}$$

9.3 THE BULK-WAVE ACOUSTO-OPTIC SPECTRUM ANALYZER

Suppose now that instead of having a single frequency component, the rf acoustic signal has a power spectral density $S(f)$. The quantity $S(f)$ simply represents the amount of acoustic power contained in a differential frequency range df. Suppose we assume that the intensity I of the deflected optical beam at any frequency f is linearly proportional to the power spectral density at this frequency, $S(f)$, and that the deflection angle θ is also a linear function of f. Then if we can measure $I(\theta)$, there must be a one-to-one correspondence between this quantity and $S(f)$. Thus the acousto-optic deflection provides a means for measuring the Fourier power spectrum of an rf signal.

The implementation of such an analyzer is shown in Figure 9.7. The device consists of the acousto-optic beam deflector just analyzed followed by a lens of focal length L_f and an array of N photodetectors placed at the focal plane.

To understand how the analyzer operates, first consider the properties of an ideal lens. From elementary physics, the reader is aware that the lens has the property of transforming radiation from an incident plane wave into a point at the lens' focal plane.

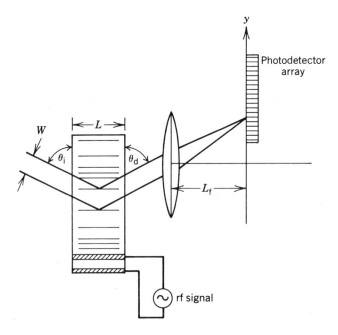

Figure 9.7 Bulk-wave spectrum analyzer.

The location y of the point in the focal plane is obtained by noting that any ray passing through the lens center travels along an unbent path. Thus, from Figure 9.8 a plane wave incident at an angle θ will be focused at the point $y = L_f \tan \theta$, which for small angles of incidence is approximated as

$$y \cong L_f \, \theta \qquad (9.7)$$

Since by definition of an ideal lens, all other rays of the plane wave must focus at the same point, these rays can easily be added to the figure as indicated by the dashed lines.

However, just as the acoustic beam must have finite width, so must the optical beam. Let us assume the optical beam width is W. A beam that is normally incident upon the lens is shown in Figure 9.9a. By our previous discussion, such a beam is equivalent to a superposition of plane waves with angular spread

$$\delta\theta \cong \frac{\lambda}{W} \qquad (9.8)$$

By ray tracing, we observe from Figure 9.9b that these plane waves will not be focused at a single point at the focal plane, but rather over a distance

$$\delta y \cong L_f \delta\theta = \frac{L_f \lambda}{W} \qquad (9.9)$$

The quantity δy is called the minimum resolvable spot size.

The operation of the acousto-optic spectrum analyzer can now be fully described.

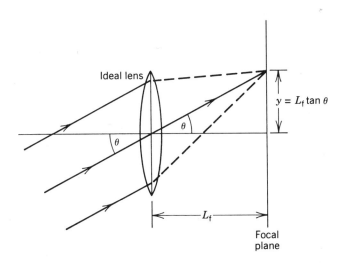

Figure 9.8 Focusing properties of an ideal lens.

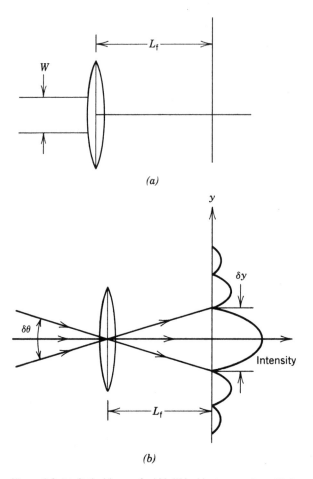

(a)

(b)

Figure 9.9 (a) Optical beam of width W incident upon a lens; (b) Intensity distribution at the focal plane.

As the frequency of the acoustic source is varied, the angular direction of the deflected beam will vary correspondingly. Because of the focusing lens, each frequency will be imaged at a different point y along the focal plane. Two different frequencies will just be resolvable when they are separated spatially by an amount nominally equal to δy. Suppose the acoustic transducer has a bandwidth equal to Δf. The total "swing" Δy_T in the deflected beam over this frequency range is obtained using Eqs. 9.5 and 9.7:

$$\Delta y_T = L_f (2\delta\phi_{max}) = L_f \frac{\Delta f \, \lambda}{v_{ac}}$$

Thus, the number of resolvable frequencies or spots N is given by

$$N = \frac{\delta y_T}{\delta y} = \frac{\Delta f W}{v_{ac}} \cong \Delta f \tau \qquad (9.10)$$

where τ is the transit time required for the acoustic wave to transverse the optical beam. Note that N determines the frequency resolution of the analyzer.

9.4 GUIDED-WAVE BEAM DEFLECTOR

9.4.1 Deflector Geometry

Let us now investigate how the Bragg effect can be applied to guided-wave structures. A typical geometry for an electro-optical waveguide modulator/deflector is shown in Figure 9.10*a*. The modulator consists of a narrow guiding electro-optic film located on a substrate surface with a thin-film metallic interdigital electrode array formed on top of the film using standard photolithographic techniques. The waveguide itself can be fabricated, for example, by the polycrystalline deposition of ZnO on a low-index substrate or by Nb metal diffusion into LiTaO$_3$.

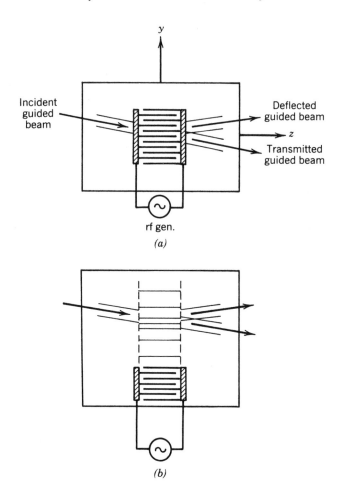

Figure 9.10 (*a*) Electro-optic beam deflector; (*b*) acousto-optic beam deflector.

Application of a time-varying voltage to the electrodes produces in this region a time-varying electric field which is spatially periodic in the $\hat{y}$ direction and which has a magnitude that decreases with distance into the film. Because the film is electro-optic, the spatial variation in electric field is mirrored by a corresponding spatial perturbation of the refractive index. This perturbation must, therefore, also have the electrode periodicity along $\hat{y}$ and an amplitude that decreases with distance into the film. Generally, the electric field-generated modification to the permittivity is quite small.

Alternatively, an optical phase grating may be generated using an acoustic wave as shown in Figure 9.10b. The waveguide consists of a substrate and guiding film at least one of which is piezoelectric. Typical substrate–film configurations include various sputtered high-refractive-index glass films above a piezoelectric quartz substrate, metal-diffused layers near the surface of a piezoelectric LiNbO$_3$ substrate, or a sputtered piezoelectric ZnO film above a fused quartz substrate.

The periodic perturbations in refractive index are produced via the excitation of a surface acoustic wave (SAW). The SAW is somewhat analogous to the guided optical wave in that it is guided along the top surface of the substrate with its stored acoustic energy decreasing exponentially into the substrate bulk. The SAW is launched by means of a set of interdigitated thin-film metal electrodes located on the surface of the pie-zoelectric substrate or guiding layer. Application of a time-varying voltage produces on each finger a corresponding time-varying charge. Through the piezoelectric effect, the charge produces a time-varying surface stress which excites a $\hat{y}$-directed propagating surface wave much in the same fashion a waves are produced by dropping a rock into a pond. By proper adjustment of the excitation frequency, the wavelength of the acoustic signal launched by each finger can be made equal to the periodic spacing between electrodes so that the acoustic signals launched from each finger all add in phase thereby producing a strong acoustic signal. This effect is discussed in greater detail in the problems. As the acoustic wave travels across the crystal, the resulting periodic compressions and rarefractions of the crystal atoms induce corresponding periodic increases and decreases in the refractive index. Because of the nature of the surface wave, the amplitude of the particle motion is strongest at the surface, decaying ex-ponentially into the substrate. Typically, the amplitude is significant over a depth comparable to the acoustic wavelength. The nature of the acoustic-wave grating differs slightly from that of the electro-optically induced type in that the grating propagates along y with the acoustic-wave velocity. However, as discussed in Section 9.1, for most cases the disturbance may be considered stationary due to the much slower velocity of the acoustic wave relative to the optical signal.

9.4.2   Coupled-Mode Equations

To analyze quantitatively the effect of the periodic perturbation, use is again made of the Lorentz reciprocity theorem. With reference to Figure 9.11, we shall treat the indicated asymmetric slab waveguide as the unperturbed geometry and the electro-optically or the acoustically induced periodic modulation of the refractive index as the perturbation.

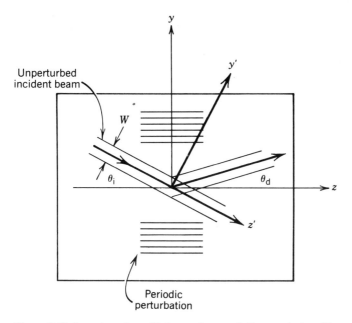

Figure 9.11 Scattering of a guided wave by a periodic perturbation. The angles θ_i and θ_d are shown exaggerated for clarity.

In absence of the acoustic disturbance, we let the unperturbed fields correspond to an incident guided wave propagating at an angle θ_i with respect to the $\hat{z}$ axis as shown. To simplify the analysis, it will be assumed that the unperturbed slab waveguide has rotational symmetry with respect to the x axis and thus the propagation constant must be independent of θ_i. In fact, for any θ_i the fields are simply those of the asymmetric waveguide analyzed in Chapter 4 but referenced, however, with respect to the primed rather than unprimed coordinate system. Thus the unperturbed fields are given by

$$\left.\begin{array}{c} \mathbf{E}(x, z') \\ \mathbf{H}(x, z') \end{array}\right\} = \left\{\begin{array}{c} \mathbf{e}_i(x) \\ \mathbf{h}_i(x) \end{array}\right\} e^{-jk_z z'}$$

Vector directions referenced
to primed coordinates

As was demonstrated in Example 9.1, the angle θ_i necessary for Bragg deflection is generally quite small. We, therefore, make the simplifying assumption that the fields $\mathbf{e}_i(x)$ and $\mathbf{h}_i(x)$ can be described to good approximation as those for a wave traveling exactly along z, that is, those fields described in Chapter 4. Transforming the variable z' into the yz coordinate system, the unperturbed fields become

$$\left.\begin{array}{c} \mathbf{E}(x, y, z) \\ \mathbf{H}(x, y, z) \end{array}\right\} \cong \left\{\begin{array}{c} \mathbf{e}(x) \\ \mathbf{h}(x) \end{array}\right\} e^{-jk_{yi} y} e^{-jk_{zi} z} \tag{9.11}$$

where

$$k_{yi} = -k_z \sin \theta_i \qquad k_{zi} = k_z \cos \theta_i$$

and $e(x)$ and $h(x)$ are the fields for the asymmetric slab waveguide described in Chapter 4. The lowercased letters indicate that we have normalized the field amplitudes to carry unity power.

In presence of the acoustic perturbation, the previous analysis of bulk-wave Bragg scattering leads us to assume that two guided modes should exist: an incident wave propagating at an angle θ_i and a deflected wave propagating at an angle θ_d. The perturbed fields must therefore be a linear combination of these two solutions, that is

$$\left. \begin{array}{c} \mathbf{E}'(x, y, z) \\ \mathbf{H}'(x, y, z) \end{array} \right\} = a_i(z) \left\{ \begin{array}{c} \mathbf{e}(x) \\ \mathbf{h}(x) \end{array} \right\} e^{-jk_{yi}y} e^{-jk_{zi}z}$$

$$+ a_d(z) \left\{ \begin{array}{c} \mathbf{e}(x) \\ \mathbf{h}(x) \end{array} \right\} e^{-jk_{yd}y} e^{-jk_{zd}z} \qquad (9.12)$$

with

$$k_{yd} = k_z \sin \theta_d \qquad k_{zd} = k_z \cos \theta_d$$

Note that the unknown amplitude coefficients are assumed to be a function of position since we anticipate that the amplitude of the incident wave should initially decrease while traversing through the periodic region, its power being scattered into the deflected wave. Equation 9.12 can be written compactly as

$$\left. \begin{array}{c} \mathbf{E}'(x, y, z) \\ \mathbf{H}'(x, y, z) \end{array} \right\} = \sum_{p=i,d} a_p(z) \left\{ \begin{array}{c} \mathbf{e}(x) \\ \mathbf{h}(x) \end{array} \right\} e^{-jk_{yp}y} e^{-jk_{zp}z} \qquad (9.13)$$

Let us now substitute the unperturbed and perturbed fields into the Lorentz reciprocity relation, Eq. 8.10. For the left side we obtain

$$\iint_S \frac{\partial}{\partial z} (\mathbf{E}_t^* \times \mathbf{H}_t' + \mathbf{E}_t' \times \mathbf{H}_t^*) \cdot \hat{z} \, ds = \sum_{p=i,d} \left[\frac{\partial}{\partial z} a_p(z) e^{jZ_{ip}z} \right] \left[W \frac{\sin(Y_{ip}W/2)}{(Y_{ip}W/2)} \right]$$

where

$$Y_{ip} = k_{yi} - k_{yp} \qquad Z_{ip} = k_{zi} - k_{zp}$$

The $\sin(\pi x)/(\pi x)$ or sinc(x) function is plotted in Figure 9.12. Note that because of the summation in Eq. 9.13, Y_{ip} takes on two values. For $p = i$, $Y_{ii} = 0$ and the sinc function is at its maximum. For $p = d$, $|Y_{id}| = k_z (\sin \theta_i + \sin \theta_d)$. Since we anticipate that $\theta_d \cong \theta_i$ for strong Bragg deflection, then $|Y_{id}W/2| \cong k_z W \sin \theta_i \cong$

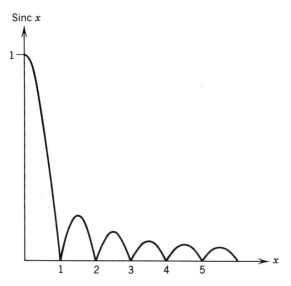

Figure 9.12 The sinc function.

$2\pi(W/\lambda) \sin \theta_i$. Therefore, provided that the incident beam width W is sufficiently large so that $W/\lambda \sin \theta_i \gg 1$, the term in the summation for which $p = d$ can be neglected. The left side of the Lorentz reciprocity theorem then is simply equal to $W(\partial/\partial z)a_i(z)$.

To evaluate the right side of the Lorentz reciprocity theorem, it is necessary to specify the functional form of the perturbation in terms of its modification to the permittivity $\Delta\epsilon$. We assume that the acoustic disturbance propagates in the $+\hat{y}$ direction with a propagation constant $\gamma = 2\pi/\Lambda$, and that the acoustic amplitude is a function of x (generally exponentially decreasing with depth into the substrate). Thus, $\Delta\epsilon$ can be described generally by

$$\Delta\epsilon(x, y) = \Delta\epsilon u(x) \cos(\gamma y)$$

where $0 < u(x) < 1$ and describes the variation of permittivity with depth into the substrate. Substitution of the above form for the perturbation along with field expressions 9.11 and 9.13 into Eq. 8.10 then yields for the right side of Eq. 8.10

$$-j \sum_{p=i,d} a_p(z)e^{jZ_{ip}z}C\left[\frac{W \sin(Y_{ip} - \gamma)(W/2)}{(Y_{ip} - \gamma)(W/2)} + \frac{W \sin(Y_{ip} + \gamma)(W/2)}{(Y_{ip} + \gamma)(W/2)}\right]$$

where

$$C = \frac{\omega\Delta\epsilon}{4} \int_{-\infty}^{\infty} dx \, u(x)|\mathbf{e}(x)|^2 \qquad (9.14)$$

By identical arguments to those made above, one of the two terms in brackets is significant if and only if

$$Y_{ip} \pm \gamma = 0 \qquad (9.15)$$

where p can take on only the values i,d. When $p = i$, $Y_{ii} = 0$ and Eq. 9.15 cannot be satisfied. Thus, for $p = i$, neither term in the brackets is significant. When $p = d$ Eq. 9.15 can be satisfied if and only if

$$k_z (\sin \theta_i + \sin \theta_d) = \pm \gamma \qquad (9.16)$$

When Eq. 9.16 is satisfied, one of the sinc terms in brackets is equal to unity and the other is small and Eq. 8.10 simplifies to

$$\frac{\partial}{\partial z} a_i(z) = -ja_d(z)Ce^{jZ_{id}z} \qquad (9.17)$$

Because the incident angle θ_i was chosen arbitrarily, we can define a new incident angle θ_d which by symmetry must have a corresponding deflection angle θ_i. This interchange of θ_i and θ_d is equivalent to changing the subscripts i and d in Eq. 9.17 yielding the second relation

$$\frac{\partial}{\partial z} a_d(z) = -ja_i(z)Ce^{jZ_{di}z} \qquad (9.18)$$

Equations 9.17 and 9.18 are the familiar coupled equations of motion analyzed in Section 8.6. If the acoustic beam has width L in the z direction, then the ratio of the deflected power at L, $P_d(L)$, to incident power at $z = 0$, $P_i(0)$, or deflection efficiency, η_{defl}, is given by

$$\eta_{defl} = \frac{P_d(L)}{P_i(0)} = \left(\frac{CL}{SL}\right)^2 \sin^2(SL) \qquad (9.19)$$

where

$$SL = \sqrt{(Z_{id}L/2)^2 + (CL)^2}$$

For strong coupling to occur, the term $Z_{id} = -Z_{di}$ must approach zero or

$$k_z(\cos \theta_i - \cos \theta_d) = 0 \qquad (9.20)$$

which requires that $\theta_i = \pm \theta_d$. However, for relation 9.16 to be simultaneously satisfied we must choose $\theta_i = +\theta_d$. In this situation, Eq. 9.16 becomes

$$2k_z \sin \theta_i = +\gamma$$

which is precisely the Bragg condition.

We have therefore not only derived the Bragg condition rigorously, but also showed the consequences of small deviations from the Bragg condition on conversion efficiency. Further, from Eq. 9.19 when the Bragg condition is satisfied, $S = C$ and

$$\eta_{\text{defl}} = \sin^2 CL \qquad (9.21)$$

Since from Eq. 9.14 the coupling coefficient C is linearly proportional to $\Delta\epsilon$, increasing the acoustic power P_{ac} should increase η_{defl}. To determine exactly how C depends on the applied acoustic power, we need to obtain an explicit relation between $\Delta\epsilon$ and P_{ac}. This relation is discussed in the following section for the acousto-optical interaction.

9.4.3   The Photoelastic Effect

Distortions in the crystal lattice brought about by the compressions and rarefractions of an acoustic wave create a perturbation in permittivity. Let us therefore first determine the relationship that exists between the acoustic-wave power P_{ac} and the resultant material strain s. Consider therefore the bar shown in Figure 9.13. Hooke's law states that the fractional change in length of the bar u/l, is linearly proportional to the pressure applied to its ends. Here u represents a small change in the bar length from its unstressed state. Thus if a force F is applied to a face of the bar having area A then

$$\frac{u}{l} = \frac{1}{c}\frac{F}{A} \equiv s \qquad (9.22)$$

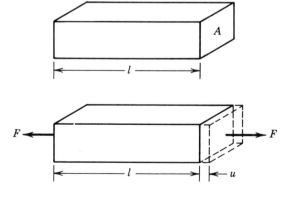

Figure 9.13 Distortion of a bar by a force along its axis.

where c is known as the elastic constant, or Young's modulus. The fractional change in length u/l is also called the strain s. Let us determine the increase in stored energy experienced by such a bar in going from an unstrained state $(u = 0)$ to some strained state $(u = u_0)$. The differential work dW required to stretch the bar by a differential amount du is

$$dW = F\,du = \frac{Ac}{l}u\,du$$

and thus the total work W required to stretch the bar a distance u_0 is

$$W = \frac{Ac}{2l}u_0^2 = (Al)\frac{c}{2}\left(\frac{u_0}{l}\right)^2$$

Making use of our definition of strain, the energy per unit volume stored by the bar is

$$\frac{W}{V} = \frac{cs_0^2}{2} \tag{9.23}$$

where

$$s_0 = u_0/l$$

Suppose now that we launch an acoustic beam of cross-sectional area A traveling with acoustic velocity v_{ac}. Figure 9.14 shows a "snapshot" of the leading edge of the wave at two times separated by Δt. During this period, all the energy stored in a volume $V = A\Delta y$ is transferred to the right of our observation plane located at $y = 0$. Thus, the rate of transfer of acoustic energy to the right, that is, the acoustic power flow, is from Eq. 9.23

$$\frac{\Delta W}{\Delta t} = P_{ac} = \frac{cs_0^2}{2}\frac{(A\Delta y)}{\Delta t}$$

or

$$P_{ac} = \frac{c}{2}s_0^2\,v_{ac}\,A \tag{9.24}$$

From Eq. 9.24 we see that the material strain is therefore related to the acoustic power by

$$s_0 \propto \sqrt{P_{ac}/A}$$

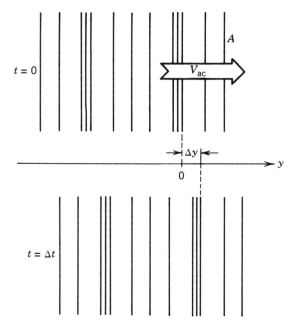

Figure 9.14 Position of acoustic wavefronts for two times separated by Δt.

If we further assume that the change in material permittivity is linearly related to the applied strain, we obtain the desired result:

$$\frac{\Delta \epsilon}{\epsilon_0} = \beta \sqrt{P_{ac}/A} \tag{9.25}$$

The constant β is dependent on the material's acoustic and so-called photoelastic properties, the latter being simply a measure of the change in optical permittivity per unit acoustic strain. The value of β is also a function of the propagation directions of the acoustic and optical waves within the material as well as the polarization of the optical signal and directions of particle motion for the acoustic disturbance.

9.4.4 Bragg Deflection Efficiency

Having related $\Delta \epsilon$ to P_{ac}, we are now in a position to compute the deflection efficiency versus acoustic power input. From Eqs. 9.14, 9.21, and 9.25, the deflection efficiency η_{defl} is

$$\eta_{defl} = \sin^2(CL)$$

where

$$CL = \frac{\omega}{4} \epsilon_0 \beta \sqrt{P_{ac}L^2/A_{eff}} \int_{-\infty}^{\infty} dx\, u(x)|e(x)|^2 \tag{9.26}$$

and the effective cross-sectional area of the beam, A_{eff}, is defined by

$$A_{eff} = L \int_{-\infty}^{\infty} dx\, u(x)$$

For strain fields which decay exponentially, then

$$u(x) = 0 \qquad x > 0$$
$$= e^{-x/h} \qquad x < 0$$

and

$$A_{eff} = Lh$$

Substitution of our expression for A_{eff} into Eq. 9.26 and multiplying and dividing by the characteristics impedance of free space then yields

$$CL = \frac{\pi}{\lambda_0} \sqrt{P_{ac}LM_{eff}/2h} \tag{9.27}$$

where the effective figure of merit M_{eff} is defined by

$$M_{eff} = \frac{\beta^2}{2} \frac{\epsilon_0}{\mu_0} \left(\int_{-\infty}^{\infty} dx\, u(x)|e(x)|^2 \right)^2 \tag{9.28}$$

Although CL must be computed for each new geometry of interest, several important points can be made, however, by examining its functional form. First, it is observed that CL is maximized by maximizing the spatial overlap of the acoustic strain $u(x)$ and the electric field $|e(x)|$. When this is accomplished, CL is further increased by reducing h, that is by "squeezing" the total acoustic power into a smaller region below the substrate surface.

When the coupling CL is small, the deflection efficiency becomes

$$\eta_{defl} \cong (CL)^2 \tag{9.29}$$

or

$$\eta_{defl} = \left(\frac{\pi}{\lambda_0} \right)^2 \left(M_{eff} \frac{P_{ac}L}{2h} \right)$$

Thus, in this limit the deflected beam intensity is linearly proportional to the acoustic power and effective figure of merit.

Implementation of guided-wave Bragg deflection for spectrum analysis is shown schematically in Figure 9.15. A semiconductor laser is butted up against one end of a dielectric waveguide and its radiated power is coupled into the film. The laser aperture is generally quite small so that the beam diverges rapidly. The beam is collimated using one of the three types of lenses shown in Figure 9.16. To understand how these lenses work we note that a lens has the property of introducing a phase shift for paraxial rays passing through it which decreases quadratically with distance away from the lens axis. Standard optical lenses achieve this effect by being widest at the center with a thickness that decreases quadratically toward the edge. For planar lenses, the same phase shift can be achieved in a number of ways. In the Luneburg lens, shown in Figure 9.16a, the phase shift is controlled by placing a thicker layer of high-refractive-index material near the center of the lens, the thickness decreasing quadratically toward the lens perimeter. The thicker central regions produce a higher effective index, a correspondingly lower velocity, and thus a larger phase shift. In the geodesic lens, a spheroidal depression is introduced into the waveguide as shown in Figure 9.16b. Rays propagating through the central lens region must travel over a longer path than those near the periphery. The associated phase shift for a ray increases quadratically with distance away from the lens axis. For this type of lens, the effective index for all rays is the same. In the Fresnel lens, shown in Figure 9.16c, use is made of the fact that it is not the absolute phase shift through the lens that matters but only its value to within a multiple of 2π. Because lenses have phase shifts that change quadratically with axial distance x, phase changes rapidly as we go off axis. The Fresnel lens discretizes this phase shift at multiples of π. This is shown diagramatically in Figure 9.17. In practice, zones of phase shift alternating by π are produced by introducing a

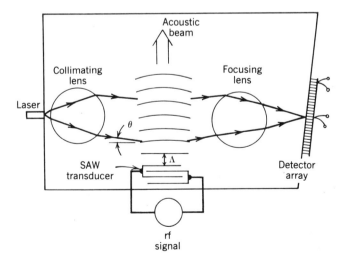

Figure 9.15 Practical implementation of an integrated optics spectrum analyzer.

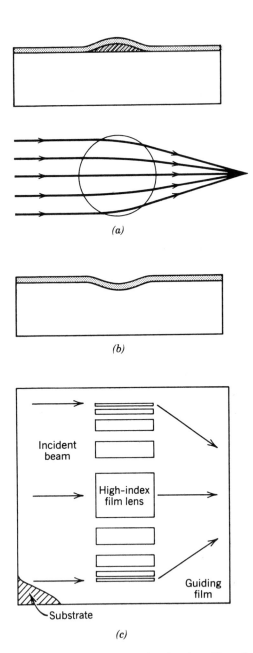

(a)

(b)

(c)

Figure 9.16 Planar lenses. (*a*) Luneburg lens; (*b*) geodesic lens; (*c*) Fresnel lens. (After Chang *et al.* (Ref. 3). Copyright 1980, IEEE. Reproduced by permission.)

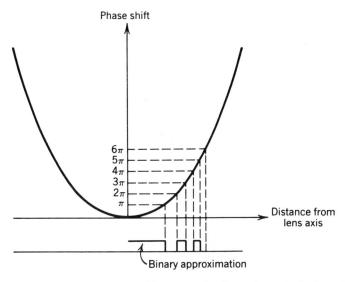

Figure 9.17 Construction of a binary approximation to the quadratic phase shift of an ideal lens. (After Chang *et al.* (Ref. 3). Copyright 1980, IEEE. Reproduced by permission.)

higher refractive-index thin film of appropriate spacings above the waveguide as shown in Figure 9.16c.

The collimated radiation from the output of the planar lens is next incident on the acoustic grating, fabricated as discussed previously. A second planar lens focuses the deflected beam onto one of a number of detectors which can be butt coupled to the output waveguide edge.

PROBLEMS

9.1 A plane acoustic wave propagates in an acousto-optic crystal with a velocity of 2500 m/s. What is the acoustic frequency required to deflect a He–Ne laser beam at a Bragg angle of 1°? The crystal has a refractive index $n = 1.6$.

9.2 What is the difference in deflection angle for a He–Ne laser beam that is Bragg scattered by an acoustic plane wave at frequencies of 50 and 75 MHz? The acoustic velocity is 3000 m/s and the crystal refractive index is 2.0.

9.3 A bulk-wave acousto-optic deflector is to operate with an acoustic center frequency of 500 MHz and bandwidth $\Delta f = 100$ MHz. The acousto-optic material has $v_{ac} = 3000$ m/s and $n = 1.5$. What is the maximum allowable transducer aperture L? Assume that a He–Ne laser is used.

9.4 What is the minimum spot size to which a 1-mm-diameter He–Ne laser beam can be focused using a lens having a 3-cm focal length?

9.5 A bulk-wave modulator is fabricated with a material having acoustic velocity $v_{ac} = 3000$ m/s and the refractive index $n = 1.4$. The modulator has a center frequency of 600 MHz. If the acoustic beam is used to deflect a He–Ne laser beam of width $W = 1$ mm, what must be the acoustic beam width L such that 100 different frequencies can be resolved? Assume that the detectors are placed at the output focal plane of a lens having $L_f = 4$ cm.

9.6   A surface-acoustic-wave (SAW) transducer consists of a series of $2N$ identical interdigitated metal fingers of alternating polarity as shown in Figure 9.18.

(a) Suppose that each of the N finger pairs located at the points $y = nd$, $n = 0$, 1, . . ., launches a forward- and backward-propagating acoustic wave having a stress field at the surface given by

$$s(y) = s_0[e^{-j\omega(y-nd)/v_{ac}} + e^{+j\omega(y-nd)/v_{ac}}]$$

Using superposition, show that the amplitude of either the total forward- or the total backward-stress wave is given by

$$|s(y)| = s_0 \frac{\sin(N\omega d/2\,v_{ac})}{\sin(\omega d/2\,v_{ac})}$$

(b) Show that $|s|$ has a maximum at the center frequency $f_0 = v_{ac}/d$ and compute its value.

(c) Show that the fractional bandwidth $\Delta f/f_0$ of the SAW transducer is equal to the reciprocal of the number of finger pairs N.

(d) Based upon your answers for parts (b) and (c) what general conclusions can be derived concerning the relationship between SAW acoustic power and bandwidth?

9.7   Figure 9.19 shows a portion of a bar under a nonuniform stress having cross-sectional area A. The force at two points located a distance Δx apart is as shown and causes the atoms located at $x + \Delta x$ and x to be displaced from their equilibrium positions by the amounts $u(x + \Delta x)$ and $u(x)$, respectively.

(a) Use Hooke's law to show that in the limit of $\Delta x \to 0$,

$$s(x) = \frac{\partial u(x)}{\partial x} = \frac{1}{c}\frac{F(x)}{A}$$

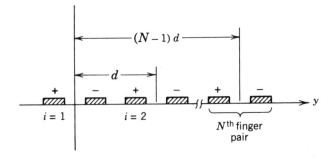

Figure 9.18 Interdigital electrode position for a surface acoustic wave (SAW) transducer.

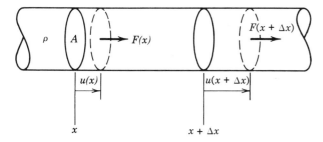

Figure 9.19 Bar under nonuniform stress.

(b) Use Newton's law, $F = ma$, in the differential element of volume located between x and $x + \Delta x$ to show that if the bar mass density is equal to ρ, and the bar forces are time varying, then

$$\frac{\partial}{\partial x}\left[\frac{F(x, t)}{A}\right] = \rho\frac{\partial}{\partial t}v(x, t)$$

where

$$v \equiv \frac{\partial}{\partial t}u(x, t)$$

(c) Use the results of part (b) and the fact that from part (a)

$$c\frac{\partial}{\partial t}\left[\frac{F(x, t)}{A}\right] = \frac{\partial}{\partial x}v(x, t)$$

to show that the acoustic-wave velocity v_{ac} is given by

$$v_{ac} = \sqrt{c/\rho}$$

9.8 The photoelastic constant p is usually defined for an isotropic material by the relation

$$\frac{d}{ds}\left(\frac{1}{n^2}\right) = p$$

where s is the material strain and n is the refractive index.

(a) Using the above definition for p and the results of problem 9.7, show from Eqs. 9.24 and 9.25 that the proportionality constant β is related to p by

$$\beta = 2n\sqrt{n^6p^2/2\rho\ v_{ac}^3}$$

(b) Show from Eq. 9.28 and the results of part (a) that for a bulk acoustic-

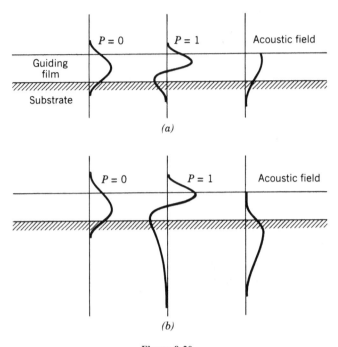

Figure 9.20

wave modulator having cross-sectional area A the effective figure of merit is simply

$$M_{\text{eff}} = \frac{n^6 p^2}{\rho v_{\text{ac}}^3}$$

9.9 An integrated optics spectrum analyzer uses a SAW transducer having an acoustic aperture of $L = 8.7$ mm with a center frequency $f_0 = 175$ MHz and velocity $v_{\text{ac}} = 3000$ m/s. It is found that 10 mW of acoustic power results in a 70% deflection efficiency for a He–Ne beam. Assuming that the acoustic beam has a $1/e$ spatial decay h equal to the acoustic wavelength Λ what is the value of M_{eff}?

9.10 A geodesic lens having a focal length L_f of 4.5 mm is located a distance L_f in front of a semiconductor laser. If the laser aperture is 6 μm and its wavelength is 1 μm, what is the beam width of the laser radiation on the far side of the lens?

9.11 (a) Figure 9.20a shows the acoustic stress field and optical field amplitude for the first two modes on a dielectric waveguide. If all other factors are the same, which waveguide mode would you anticipate to have the strongest Bragg deflection? Why?

  (b) Repeat for Figure 9.20b.

REFERENCES

1. Anderson, D. B., *et al.* "Comparison of optical-waveguide lens technologies." *IEEE Journal of Quantum Electronics* QE-13 (1977): 275–282.
2. Chang, W. S. C., "Acoustooptical deflections in thin films." *IEEE Journal of Quantum Electronics* QE-7 (1971): 167–170.
3. Chang, W. S. C., and Ashley, P. R. "Fresnel lenses in optical waveguides." *IEEE Journal of Quantum Electronics* QE-16 (1980): 744–747.
4. Chang, I. C. "Acoustooptic devices and applications, Part I." *IEEE Transactions on Sonics and Ultrasoncis* SU-23 (1976): 2–22.
5. Gordon, E. I. "A review of acoustooptical deflection and modulation devices." *Applied Optics* 5 (1966): 1629–1639.
6. Korpel, A. "Acousto-optics—A review of fundamentals." *Proceedings of the IEEE* 69 (1981): 48–53.
7. Lean, E. G. H., White, J. M., and Wilkinson, C. D. W. "Thin-film acoustooptic devices." *Proceedings of the IEEE* 64 (1976): 779–788.
8. Pinnow, D. A. "Guidelines for the selection of acoustooptic materials." *IEEE Journal of Quantum Electronics* QE-6 (1970): 223–238.
9. Schmidt, R. V. "Acoustooptic interactions between guided optical waves and acoustic surface waves, Part II." *IEEE Transactions on Sonics and Ultrasonics* SU-23 (1976): 22–33.
10. Schmidt, R. V., and Kaminow, I. P. "Acoustooptic Bragg deflection in $LiNbO_3$ Ti-diffused waveguides." *IEEE Journal of Quantum Electronics* Correspondence (Jan. 1975): 57–59.
11. Valette, S., *et al.* "Integrated optical spectrum analyser using planar technology on oxidised silicon substrate." *Electronics Letters* 19 (1983): 883–885.

CHAPTER 10

Optical Fibers

10.1 INTRODUCTION

Much of the research and development in the area of integrated optics has been stimulated by technological breakthroughs in the production of low-loss optical fibers for telecommunication applications. Whereas the best existing optical fibers in 1966 had propagation losses in excess of 1000 dB/km, such attenuation has now been reduced to as low as 0.2 dB/km for source wavelengths near 1.55 μm. The combination of low loss and wide bandwidth available at optical frequencies make these fibers extremely attractive for the transmission medium in communication systems. These benefits are clearly demonstrated in Figure 10.1 which compares attenuation versus carrier frequency for a single-mode silica fiber operated at 1.55 μm and three commonly used metallic guiding media. Whereas the metallic transmission lines suffer from exponentially increasing conduction losses at higher operating frequency (and bandwidth), the dielectric fiber exhibits no such losses. For signal frequencies above several kilohertz, the losses of the fiber line are far less, thereby allowing repeaters (amplifiers) to be spaced farther apart, with a subsequent savings in cost. As an example, fiber communication systems have been demonstrated that can transmit information at a rate of over two billion bits per second over 130 km without a repeater with an error rate of 1 per billion bits. At this information and error rate the text of five entire sets of a 30-volume encyclopedia could be transferred between New York and Philadelphia within a second and the only fault might be that two letters in the text would be capitalized instead of lowercased.

Additionally, optical fibers enjoy a number of other advantages over metallic transmission lines. Because the fibers are made from a dielectric and therefore the link between source and detector is nonmetallic, they avoid ground loop pick-up problems resulting from electromagnetic interference. Such immunity is important for military applications since it permits cables for highly sensitive electronics to be located near electrical power system wiring in land, sea, and aircraft.

The small size and light weight of optical fibers also makes them desirable for replacing conventional wiring in aircraft where space and weight are at a premium. Fibers can also be used to replace metal cables in existing underground wiring ducts and in buildings, thereby increasing available user bandwidth for phone and data communication systems.

10.2 TYPES OF FIBER GEOMETRIES

The optical fiber shown in Figure 10.2 is a cylindrical structure consisting of a central core of radius a and permittivity ϵ_2 surrounded by a concentric cladding region of

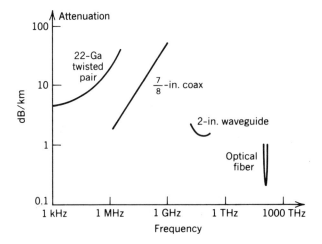

Figure 10.1 Attenuation of various guiding media. (After Henry (Ref. 8). Copyright 1985, IEEE. Reproduced by permission.)

slightly lower dielectric constant ϵ_1. Lowest loss fibers are generally fabricated of a silica (SiO_2) core doped with either GeO_2 or P_2O_5 to increase the refractive index and are surrounded by a silica- or P_2O_5-doped silica cladding. For applications in which attenuation is not of primary concern, such as for short-distance communication and data links, higher loss plastic-clad or totally plastic fibers can be used.

In theory, guidance along such fibers is possible without the introduction of the cladding. In its absence, confinement would occur through total internal reflection at the core–air interface in much the same fashion as for slab dielectric waveguides. The cladding, however, serves several useful purposes. It protects the core from external surface contaminants, improves mechanical strength and reduces scattering losses resulting from dielectric discontinuities at the core surface.

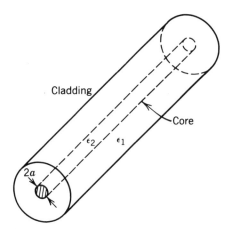

Figure 10.2 Step-index optical fiber.

Further, by varying the value of the refractive index of the cladding relative to that of the core, an additional degree of freedom in controlling waveguide propagation characteristics is obtained. Frequently, one or more additional plastic coating layers surround the cladding to provide further cable strength as well as mechanical protection from external shocks and perturbations of the fiber geometry.

A number of different fiber geometries are presently used for various communications applications. These configurations may be classified according to the variation in dielectric constant, or equivalently, refractive index along the radial direction of the fiber as shown in Figure 10.3. Waveguides having a uniform refractive index throughout but exhibiting an abrupt step at the core–cladding interface are referred to as step-index fibers. Fibers for which refractive index varies in some continuous fashion as a function of radial distance are known as graded-index fibers.

All of the above fiber geometries can be divided into two classes, monomode and multimode. In analogy with the slab guide, monomode fibers will support only one propagating mode while multimode fibers will generally support hundreds. Some typical numbers for waveguide dimensions and refractive-index variation are shown in Figure 10.3 for both types of fibers.

As we shall see in the analysis to follow, monomode fibers offer advantages over multimode fibers in terms of reduced pulse distortion for long-distance, high data rate communication systems. For shorter distance or lower bandwidth applications, however, multimode fibers have a number of advantages. The larger core diameter makes it easier to couple power into the fiber and makes the alignment for splices less critical. Further, multimode fibers can be used with light emitting diode (LED) sources which

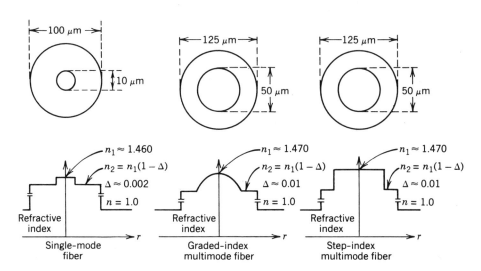

Figure 10.3 Refractive-index profile for three classes of optical fibers. (After A. H. Cherin, *An Introduction to Optical Fibers*. McGraw-Hill, New York. Copyright © 1983, Bell Telephone Laboratories, Inc. Reproduced by permission.)

are much cheaper and less complex than the laser diodes required for monomode fiber applications.

10.3 MONOMODE FIBERS

As was indicated in Section 10.2, monomode fibers are advantageous for high data rate, long-distance telecommunication systems because they do not suffer from the pulse broadening associated with multimode dispersion. Thus, the waveguiding properties of such fibers are extremely important. In this section, a powerful approximate method for solving the waveguide characteristics of these modes on both step and graded-profile fibers is introduced. The simple functional form of the solution obtained here for the step-index fiber can be contrasted with an exact solution given in terms of more complicated functions in the advanced material in Section 10.4.

10.3.1 Wave Equation for Weakly Guided Fibers

Let us consider wave propagation along the cylindrical fiber shown in Figure 10.4. The refractive index profile $n(r)$ is assumed to vary in some fashion from a peak value of n_{co} at the center of the fiber to a value of n_{cl} in the cladding. The cladding region is assumed to extend to infinity. The core–cladding interface is at $r = a$ as shown.

From the analysis of the slab dielectric waveguide, we anticipate that the propagation constant for the fundamental mode must be somewhere between the following values:

$$\frac{\omega}{c} n_{cl} < k_z < \frac{\omega}{c} n_{co}$$

The lower and upper bounds represent the low-frequency (cutoff) and high-frequency limits, respectively.

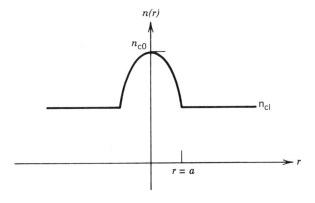

Figure 10.4 Weakly guiding graded-index fiber.

Because the fiber is to be monomode, all other modes must be cutoff. This is accomplished by making n_{co} close to n_{cl} as was indicated in Figure 10.3. Such a fiber is said to be weakly guiding. Because the fiber is weakly guiding, the evanescent fields extend very far outside the core region and the bounce angle of the wave inside the fiber is very close to 90° (see homework problem 10.1).

We see, therefore, that the fundamental mode must be nearly a plane wave, the simplest example being a wave linearly polarized in one direction, say $\hat{y}$. The fields are therefore given approximately by

$$E_y = \psi e^{-jk_z z}$$

$$H_x = -\frac{1}{\eta} E_y \tag{10.1}$$

Here ψ is a scalar representing the variation in field amplitude along the fiber radial direction and the characteristic impedance η is given by $\eta = \sqrt{\mu_0/\epsilon}$ with $\epsilon \cong \epsilon_0 n^2$. Further, because $n_{co} \cong n_{cl}$, the field properties are only weakly influenced by polarization properties of the fiber. This is readily understood by examination, for example, of the guided-wave solution for TE and TM modes on the symmetric slab dielectric waveguide. Both the guidance conditions as well as the field solutions for these two polarizations approach each other as the difference in refractive index between core and surrounding medium is reduced to zero. Accordingly, the spatial dependence of the scalar ψ must be polarization insensitive and must satisfy the scalar equivalent of the wave equation

$$\nabla_t^2 \psi - k_z^2 \psi = -k^2(r) \psi \tag{10.2}$$

where

$$k^2(r) \equiv \omega^2 \mu_0 \epsilon(r)$$

and $\epsilon(r)$ is an arbitrary function of the radial coordinate r. In the following section we show how an accurate approximate solution to this equation can be obtained.

10.3.2 Stationary Formulation

A stationary formulation is a mathematical method for obtaining an accurate estimate for some scalar parameter of interest based upon an estimation of the functional form of some quantity related to this parameter. For the problem at hand, we would like to obtain an accurate estimate for the propagation constant k_z based upon a guess of the functional form of the transverse field variation $\psi(r)$.

Perhaps the simplest method for obtaining such a relation is to multiply both sides of Eq. 10.2 by ψ and integrate over an infinite cross section perpendicular to the z

axis. This yields

$$k_z^2 = \frac{\iint dA(\psi \, \nabla_t^2 \, \psi \, + \, k^2 \, \psi^2)}{\iint dA \, \psi^2} \tag{10.3}$$

This expression can be rewritten slightly by making use of Green's theorem which is derived in Appendix 3:

$$\iint dA(\psi \, \nabla_t^2 \, \phi) \, = \, - \iint dA \, (\nabla_t \psi) \cdot (\nabla_t \phi) \tag{10.4}$$

The above identity is valid provided that the function ψ goes to zero as r approaches infinity. Substituting the above expression into Eq. 10.3 with $\phi = \psi$ yields

$$k_z^2 = \frac{\iint dA[-(\nabla_t \psi)^2 \, + \, k^2 \psi^2]}{\iint dA \, \psi^2} \tag{10.5}$$

Since Eq. 10.5 is an identity, if the exact field variation, ψ_0, is known and inserted into the right side, the exact value for k_z^2, k_{z0}^2, will be obtained.

Suppose, however, that a guess for the field behavior of ψ is substituted into the right side of Eq. 10.5. This approximate field can be represented as the sum of the true field plus an error term. That is,

$$\psi \, = \, \psi_0 \, + \, p\psi_1 \tag{10.6}$$

The parameter p is a scalar number that we are free to vary between zero and infinity. When $p = 0$ our guess for the field behavior of ψ is exact. As p is increased our guess becomes progressively worse. How "bad" the guess is obviously depends on the functional form of ψ_1 as well as the magnitude of p. For any given error function ψ_1 we can think of the value of k_z^2 calculated from Eq. 10.5 as a function of p. This functional dependence will generally look as shown in Figure 10.5. Obviously the exact functional form of $k_z^2(p)$ depends on the choice of ψ_1. The important point to notice, however, is that as p approaches zero, that is, as our trial function approaches the exact function, $k_z^2(p)$ approaches k_{z0}^2 in a linear fashion. Thus, for p small but finite, the error in k_z^2 is a linear function of p.

Suppose, however, that we could construct an expression for $k_z^2(p)$ in which k_{z0}^2 represented an absolute minimum (or maximum) with respect to p for any choice of ψ_1. This is shown diagrammatically in Figure 10.6. Such a formulation would be very advantageous for "good" guesses in which p is small since $k_z^2(p)$ is now proportional to p^2 rather than p. Such an expression is called a stationary formulation. Let us show that our expression, Eq. 10.5, for k_z^2 is in fact a stationary one.

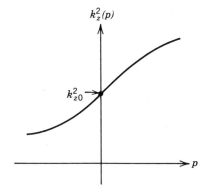

Figure 10.5 Functional dependence of k_z^2 on the error parameter p for a nonstationary formulation. Note that k_z^2 approaches the correct value k_{z0}^2 in a linear fashion with decreasing p.

Expression 10.5 can be written as

$$k_z^2(p) = \frac{N(p)}{D(p)} \tag{10.7}$$

where $N(p)$ and $D(p)$ represent the numerator and denominator of Eq. 10.5, respectively. They are functions of p because by assumption the trial fields are given by Eq.

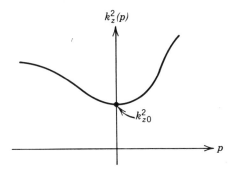

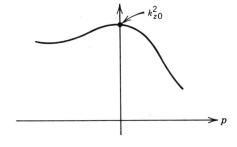

Figure 10.6 Functional dependence of k_z^2 on the error parameter p for a stationary formulation. Note that k_z^2 approaches the correct value k_{z0}^2 in a quadratic fashion with decreasing p.

10.6. For $k_z^2(p)$ to be stationary it is required that

$$\left.\frac{dk_z^2(p)}{dp}\right|_{p=0} = 0$$

or

$$D(0) N'(0) - D'(0) N(0) = 0$$

where the primes indicate differentiation with respect to p. Now from Eqs. 10.5 and 10.6 we have

$$N(p) = \iint dA[(-\nabla_t\psi_0)^2 + k^2\psi_0^2]$$

$$+ 2p \iint dA[-(\nabla_t\psi_0) \cdot (\nabla_t\psi_1) + k^2\psi_0\psi_1] + O(p^2)$$

$$D(p) = \iint dA\,\psi_0^2 + 2p \iint dA\,\psi_0\psi_1 + O(p^2)$$

where $O(p^2)$ indicates terms proportional to p^2.

Using the identity given in Eq. 10.4, $N(p)$ can be written as

$$N(p) = \iint dA\,\psi_0(\nabla_t^2\,\psi_0 + k^2\psi_0) + 2p \iint dA\,\psi_1\,(\nabla_t^2\,\psi_0 + k^2\psi_0) + O(p^2)$$

But since the true field ψ_0 satisfies wave equation 10.2, the terms in parentheses are simply equal to $k_{z0}^2\psi_0$ and $N(p)$, therefore, simplifies to

$$N(p) = k_{z0}^2 \iint dA\,\psi_0^2 + 2pk_{z0}^2 \iint dA\,\psi_0\psi_1 + O(p^2)$$

Therefore we find

$$D(0)N'(0) - D'(0)N(0)$$

$$= \left(\iint dA\,\psi_0^2\right) 2k_{z0}^2 \left(\iint dA\,\psi_0\psi_1\right) - 2\left(\iint dA\,\psi_0\psi_1\right) k_{z0}^2 \left(\iint dA\,\psi_0^2\right)$$

$$= 0$$

Thus, expression 10.5 is stationary for all ψ_1.

Let us express Eq. 10.5 in cylindrical coordinates, which are most appropriate for the cylindrical fiber. For fields that are rotationally symmetric, that is, independent of ϕ we have

$$\nabla_t\psi = \left(\frac{\partial\psi}{\partial r}\right)\hat{r}$$

and the differential element of area dA is

$$dA = r \, dr \, d\phi$$

Since all field quantities are independent of ϕ the integration over ϕ can be eliminated yielding

$$k_z^2 = \frac{\int_0^\infty r \, dr \left[-\left(\frac{\partial}{\partial r} \psi \right)^2 + k^2(r)\psi^2 \right]}{\int_0^\infty r \, dr \, \psi^2} \tag{10.8}$$

Care must be taken in using expression 10.8 to choose trial fields that vanish as r approaches infinity.

10.3.3   Solution to the Step-Index Profile

For the step-index profile, the variation in dielectric constant with radial coordinate r is

$$\epsilon(r) = \epsilon_2 = n_2^2 \, \epsilon_0 \qquad r \leq a$$

$$= \epsilon_1 = n_1^2 \, \epsilon_0 \qquad r > a$$

where a is the radius of the fiber core and n_2 and n_1 represent the core and cladding refractive indexes, respectively.

The corresponding variation in $k^2(r)$ is therefore given by

$$k^2(r) = k_0^2 n_2^2 = k_2^2 \qquad r \leq a$$

$$= k_0^2 n_1^2 = k_1^2 \qquad r > a$$

where

$$k_0^2 = \frac{\omega^2}{c^2}$$

To solve Eq. 10.8 an assumption needs to be made concerning the functional form of the field distribution ψ. Intuitively, from our analysis of the symmetric slab waveguide, it is anticipated that the field will be a maximum at the core center and decay exponentially for r sufficiently large as shown in Figure 10.7. For simplicity, let us approximate such a behavior with a continuous function of r. Such a simple function having these properties is the Gaussian distribution given by

$$\psi(r) = e^{-1/2(r/r_0)^2} \tag{10.9}$$

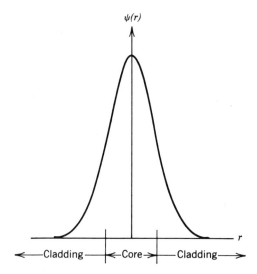

Figure 10.7 Assumed form for guided-mode field.

where r_0 represents the spot size of the mode which, at this point, is arbitrary.

Now we have already shown that a stationary formulation is one for which the best guess for ψ, that is, the true field ψ_0, gives the minimum (or maximum) value for k_z^2. Therefore, any other guess must give a correspondingly larger (or smaller) value. Let us, therefore, choose r_0 in such a manner as to minimize (or maximize) the value of k_z^2. This will then lead to the best estimate obtainable for $\psi(r)$ based upon the assumed Gaussian form of our guess. That is, we require

$$\frac{\partial k_z^2}{\partial r_0} = 0$$

This technique is referred to as a variational method.

We are now ready to evaluate the propagation constant k_z^2. Substituting the functional form for ψ given by Eq. 10.9 into Eq. 10.8 yields for the denominator

$$\int_0^\infty r\, dr\, \psi^2 = \int_0^\infty r\, dr\, e^{-(r/r_0)^2}$$

$$= \left. -\frac{r_0^2}{2} e^{-(r/r_0)^2} \right|_0^\infty$$

$$= \frac{r_0^2}{2}$$

For the numerator we obtain

$$\int_0^\infty r\,dr\,k^2(r)\psi^2$$

$$= k_2^2 \int_0^a r\,dr\,e^{-(r/r_0)^2} + k_1^2 \int_a^\infty r\,dr\,e^{-(r/r_0)^2}$$

$$= -\,(r_0^2/2)\,[e^{-(a/r_0)^2}\,(k_2^2 - k_1^2) - k_2^2]$$

and

$$\int_0^\infty r\,dr\,(\partial\psi/\partial r)^2$$

$$= \int_0^\infty \left(\frac{dr}{r_0}\right)\left(\frac{r}{r_0}\right)^3 e^{-(r/r_0)^2}$$

$$= \frac{1}{2}\int_0^\infty dx\,xe^{-x}$$

where

$$x \equiv (r/r_0)^2$$

The last expression can be integrated by parts to yield

$$\frac{1}{2}\left[\left. -xe^{-x}\right|_0^\infty + \int_0^\infty dx\,e^{-x}\right] = \frac{1}{2}$$

Thus, the expression for k_z^2 is given by

$$k_z^2 = -\frac{1}{r_0^2} - k_2^2\left[e^{-(a/r_0)^2}\frac{\Delta^2}{n_2^2} - 1\right] \tag{10.10}$$

where

$$\Delta^2 = n_2^2 - n_1^2$$

We now take the partial with respect to r_0 in the above expression and set it equal to zero to obtain the optimal value for r_0. This yields straightforwardly

$$r_0^2 = \frac{a^2}{\ln(k_0 a\Delta)^2} = \frac{a^2}{\ln v^2} \tag{10.11}$$

where, in analogy to the slab dielectric waveguide we have defined the normalized core thickness parameter

$$v = k_0 a \Delta \tag{10.12}$$

Note that we must require that $v > 1$ for r_0^2 to be defined. We shall show that this constraint is not significant for fibers that are useful in telecommunications applications.

Substituting the optimal value for r_0 back into Eq. 10.10 we obtain for the optimal estimate of k_z^2

$$k_z^2 = k_2^2 \left[1 - \left(\frac{\Delta}{n_2} \right)^2 \frac{1}{v^2} (1 + \ln v^2) \right] \tag{10.13}$$

For the weakly guiding case the quantity $(\Delta/n_2)^2$ is usually much less than 1 and k_z can be well approximated by

$$k_z \cong k_2 \left[1 - \frac{\Delta n}{n} \frac{1}{v^2}(1 + \ln v^2) \right] \tag{10.14}$$

where the fractional difference in refractive index between core and cladding is defined by

$$\frac{\Delta n}{n} \equiv \frac{n_2 - n_1}{n_2}$$

Example 10.1

Compute the effective index for a monomode fiber having a core radius $a = 3$ μm, core refractive index $n_2 = 1.460$, cladding index $n_1 = 1.455$. The fiber is excited by a source having a free-space wavelength $\lambda_0 = 1.3$ μm.

Solution

From Eq. 10.14,

$$n_{\text{eff}} \cong n_2 \left[1 - \frac{\Delta n}{n} \frac{1}{v^2}(1 + \ln v^2) \right]$$

Now

$$v \cong \frac{2\pi}{\lambda_0} a \sqrt{2 \, \Delta n \, n}$$

$$= 2\pi \frac{(3)}{1.3} \sqrt{2(0.005) \, 1.46}$$

$$= 1.75$$

Thus,

$$n_{\text{eff}} = 1.46 \left\{ 1 - \frac{0.005}{1.46} \frac{1}{(1.75)^2} [1 + \ln(1.75)^2] \right\}$$

$$= 1.457$$

10.3.4   Spatial Distribution of Light Intensity

The distribution of light intensity with r is proportional to Poynting power density $S(r)$. From Eqs. 10.1 and 10.9 we have

$$S(r) = -\tfrac{1}{2} E_y H_x^* = \frac{1}{2\eta} e^{-(r/r_0)^2} \tag{10.15}$$

The total power P carried by the guide is found by integrating the above expression over an infinite cross section yielding

$$P = \frac{1}{2\eta} \int_0^{2\pi} d\phi \int_0^\infty r \, dr \, e^{-(r/r_0)^2} = \frac{\pi}{2\eta} r_0^2$$

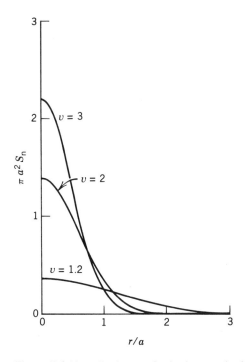

Figure 10.8 Normalized power density for several value of the parameter v.

Using the value for r_0 given by Eq. 10.11 we have for the normalized power density S_n

$$S_n(r) = \frac{\ln v^2}{\pi a^2} e^{-(r/a)^2 \ln v^2} \tag{10.16}$$

A plot of $\pi a^2 S_n$ versus r is shown in Figure 10.8 for several values of v. Note that, as stated in Section 10.3.3, for v very close to 1 the optical power is very weakly guided. Such weak guidance is undesirable because it results in high radiation losses associated with fiber bends and geometric imperfections as well as high attenuation due to dissipation within the cladding region.

10.4 MULTIMODE STEP-INDEX FIBER (ADVANCED SECTION)

In analogy to the slab dielectric waveguide, when the normalized thickness v of the step-index fiber is sufficiently large, it is expected that more than one mode may propagate. We examine the properties of these higher order modes in this section, again under the assumption that the relative difference in refractive index between the core and cladding is small. As was discussed in the last section, the small difference in refractive index implies that the solutions obtained should be nearly independent of whether the polarization of the electric field is chosen along $\hat{x}$ or $\hat{y}$. Thus, for a $\hat{y}$-polarized solution we are tempted to express our fields in the same fashion as for the monomode fiber:

$$\left.\begin{array}{c}\mathbf{E}(x, y, z) \\ \mathbf{H}(x, y, z)\end{array}\right\} \cong \left\{\begin{array}{c}\hat{y}\ \Psi(x, y) \\ \dfrac{-\hat{x}}{\eta}\ \Psi(x, y)\end{array}\right\} e^{-jk_z z} \tag{10.17}$$

where $\eta = \sqrt{\mu/(\epsilon_0 n)}$ and $n_2 \cong n_1 = n$. A little extra care must be taken, however, for the higher order modes. As we have seen from the slab waveguide, the higher order modes have fields that can vary quite rapidly with respect to the transverse coordinates. The implications of this variation can be observed by considering Maxwell's curl equations which yield under the approximation $E_x = H_y = 0$ the following expressions for the longitudinal fields E_z and H_z:

$$E_z = \frac{-j}{\eta \omega \epsilon} \frac{\partial}{\partial y} \Psi(x, y) \tag{10.18}$$

$$H_z = \frac{j}{\omega \mu} \frac{\partial}{\partial x} \Psi(x, z) \tag{10.19}$$

Thus, rapid variations in Ψ with respect to x and y can give rise to nonzero longitudinal field components E_z and H_z. Equations 10.18 and 10.19 imply that if Ψ is known then the longitudinal components are finite and can be computed.

10.4.1 Wave Equation

The functional form for Ψ can be obtained from the wave equation derived in Chapter 2 which is repeated below for convenience

$$\left(\frac{\partial^2}{\partial x^2} + \frac{\partial^2}{\partial y^2} + \omega^2\mu\epsilon - k_z^2\right)\Psi(x, y) = 0$$

Expressed in cylindrical coordinates we have

$$\left(\frac{\partial^2}{\partial r^2} + \frac{1}{r}\frac{\partial}{\partial r} + \frac{1}{r^2}\frac{\partial^2}{\partial\phi^2} + \omega^2\mu\epsilon_i - k_z^2\right)\Psi(r, \phi) = 0 \tag{10.20}$$

Note that the dielectric constant is subscripted to indicate that Eq. 10.20 represents two equations, one in the core region where $\epsilon_i = \epsilon_2$ and one in the cladding where $\epsilon_i = \epsilon_1$. Each region will have a corresponding solution for Ψ.

Using the separation of variables method we assume that $\Psi(r,\phi)$ can be expressed as

$$\Psi(r, \phi) = \psi(r)\begin{pmatrix} \cos n\phi \\ \sin n\phi \end{pmatrix} \tag{10.21}$$

where we are free to choose either the cosine or sine dependence on ϕ. Substituting Eq. 10.21 into Eq. 10.20 the wave equation for ψ becomes

$$\left[\frac{\partial^2}{\partial r^2} + \frac{1}{r}\frac{\partial}{\partial r} + (\omega^2\mu\epsilon_i - k_z^2) - \frac{n^2}{r^2}\right]\psi(r) = 0 \tag{10.22}$$

Note that because our solutions must be invariant with respect to a change in ϕ of any multiple of 2π, the quantity n defined in Eq. 10.21 must be an integer.

Equation 10.22 must now be solved for the core and cladding regions. Since it is a second-order differential equation there must be two solutions in each region. Let us define the parameter

$$k_\rho^2 = \omega^2\mu\epsilon_i - k_z^2 \tag{10.23}$$

The solutions to Eq. 10.22 are then given by the so-called Bessel functions of the first and second kind, $J_n(k_\rho r)$ and $N_n(k_\rho r)$, respectively. Any solution for $\psi(r)$ must, in general, be a linear combination of these two types of functions. The asymptotic behavior for J_n and N_n for small and large values of their arguments is given in Table 10.1. Also, shown in Figure 10.9 is a plot of the variation of these two functions with respect to their argument for several values of n.

To determine the appropriate linear combinations of these functions, their behavior is examined in the core and cladding regions.

Table 10.1 Asymptotic Values for $J_n(z)$ and $N_n(z)$ as $z \to 0$ and $z \to \infty$.

| | $z \to 0$ | | $z \to \infty$ |
	$n = 0$	$n > 0$	
$J_n(z)$	1	$\dfrac{(z/2)^n}{n!}$	$\sqrt{2/\pi z}\,\cos\left(z - \dfrac{n\pi}{2} - \dfrac{\pi}{4}\right)$
$N_n(z)$	$\left(\dfrac{2}{\pi}\right)\ln(z)$	$\dfrac{-(n-1)!}{\pi}\left(\dfrac{2}{z}\right)^n$	$\sqrt{2/\pi z}\,\sin\left(z - \dfrac{n\pi}{2} - \dfrac{\pi}{4}\right)$

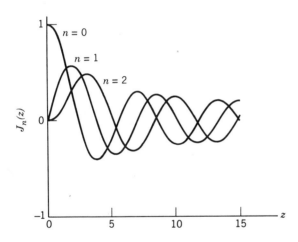

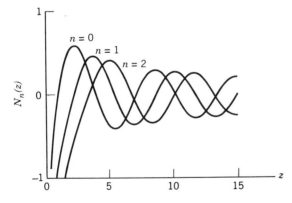

Figure 10.9 Functional behavior of $J_n(z)$ and $N_n(z)$ for several value of n.

Core Region

In analogy with the slab dielectric waveguide, it is anticipated that the propagation wavenumber k_z must be restricted between the limits

$$k_1 < k_z < k_2 \qquad (10.24)$$

where the upper and lower limits correspond to the high-frequency asymptote and cutoff frequency, respectively. Thus, in the core region it is observed from Eqs. 10.23 and 10.24 that the quantity k_ρ is always real. Defining the normalized wavenumber

$$u^2 = (k_2^2 - k_z^2)a^2$$

then

$$\psi(r) = AJ_n\left(\frac{u}{a}r\right) + BN_n\left(\frac{u}{a}r\right) \qquad r \le a$$

From Figure 10.9 and Table 10.1 it is clear that since the origin $r = 0$ is included in the core region, the function N_n cannot be used to represent $\psi(r)$ because of its divergent nature. Thus, for $r \le a$, $\psi(r) \propto J_n[(u/a)r]$ and Ψ is therefore given by

$$\Psi(r, \phi) = AJ_n\left(\frac{u}{a}r\right)\left(\begin{array}{c}\cos n\phi \\ \sin n\phi\end{array}\right)e^{-jk_z z} \qquad r \le a \qquad (10.25)$$

where the constant A is yet to be determined.

Cladding Region

From the restriction on the range of k_z given by Eq. 10.24 it is observed from the definition of k_ρ that in the cladding region k_ρ^2 is negative and k_ρ is therefore purely imaginary. Let us therefore define the normalized wavenumber

$$w^2 = (k_z^2 - \omega^2\mu\epsilon_1)a^2$$

so that the general solution for $\psi(r)$ in the cladding region is given by

$$\psi(r) = CJ_n\left(\frac{jw}{a}r\right) + DN_n\left(\frac{jw}{a}r\right)$$

For large values of r, Table 10.1 gives for the asymptotic values of J_n and N_n the expressions

$$J_n(jz) = \sqrt{1/2\pi\, jz}\,[e^{-z}e^{-j(n\pi/2 + \pi/4)} + e^{+z}e^{+j(n\pi/2 + \pi/4)}]$$

and

$$N_n(jz) = -j\sqrt{1/2\pi\,jz}\,[e^{-z}e^{-j(n\pi/2 + \pi/4)} - e^{+z}e^{+j(n\pi/2 + \pi/4)}]$$

Note that as r, or equivalently z, approaches infinity both J_n and N_n diverge due to the exponential term e^{+z}. Any linear combination involving these functions must therefore be chosen in such manner as to eliminate this behavior. This may be accomplished only by taking the linear combination

$$K_n(z) \propto J_n(jz) + j\,N_n(jz)$$

which for large z decays exponentially as z approaches infinity. The function K_n is known as the modified Bessel function of the first kind. It is plotted as a function of its argument for two values of n in Figure 10.10. Thus, we conclude that in the cladding region the field $\psi(r) \propto K_n[(w/a)r]$ and thus Ψ is given by

$$\Psi(r, \phi) = BK_n\left(\frac{w}{a}r\right)\left(\begin{matrix}\cos n\phi \\ \sin n\phi\end{matrix}\right)e^{-jk_z z} \qquad r > a \qquad (10.26)$$

where B must be determined. It is important to observe from Figures 10.9 and 10.10 that the transverse electric and magnetic fields vary in an oscillatory fashion with respect to r within the core and decay exponentially away from the core for r sufficiently large. This is precisely the same general type of behavior as was exhibited by modes on the slab dielectric waveguide.

To complete the field analysis, the longitudinal components, E_z and H_z, are computed in each region using Eqs. 10.18 and 10.19. This yields

$$E_z = \frac{-j}{\omega\epsilon_i\eta}\frac{\partial}{\partial y}\Psi(r, \phi) \qquad (10.27)$$

$$H_z = \frac{j}{\omega\mu}\frac{\partial}{\partial x}\Psi(r, \phi) \qquad (10.28)$$

The operators $\partial/\partial x$ and $\partial/\partial y$ can be expressed in cylindrical coordinates as

$$\frac{\partial}{\partial x} = \frac{dr}{dx}\frac{\partial}{\partial r} + \frac{d\phi}{dx}\frac{\partial}{\partial \phi}$$

$$\frac{\partial}{\partial y} = \frac{dr}{dy}\frac{\partial}{\partial r} + \frac{d\phi}{dy}\frac{\partial}{\partial \phi}$$

Using the transformation between rectangular and cylindrical coordinates,

$$r = \sqrt{x^2 + y^2} \qquad x = r\cos\phi$$

$$\phi = \tan^{-1}(y/x) \qquad y = r\sin\phi$$

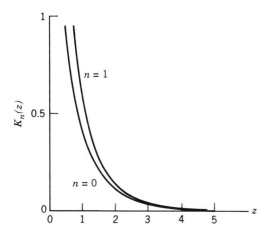

Figure 10.10 Functional behavior of $K_n(z)$ for $n = 0$ and $n = 1$.

we obtain

$$\frac{\partial}{\partial x} = \cos \phi \, \frac{\partial}{\partial r} - \frac{1}{r} \sin \phi \, \frac{\partial}{\partial \phi} \tag{10.29}$$

$$\frac{\partial}{\partial y} = \sin \phi \, \frac{\partial}{\partial r} + \frac{1}{r} \cos \phi \, \frac{\partial}{\partial \phi} \tag{10.30}$$

Let us first compute E_z. Using Eqs. 10.25, 10.27, and 10.30 yields for $r \le a$

$$E_z = \frac{-jA}{\omega \epsilon_2 \eta} \left[\frac{u}{a} J_n'\left(\frac{u}{a} r\right) \binom{\cos n\phi}{\sin n\phi} \sin \phi + \frac{n}{r} J_n\left(\frac{u}{a} r\right) \binom{-\sin n\phi}{\cos n\phi} \cos \phi \right] \tag{10.31}$$

where the prime in the above expression denotes differentiation with respect to the argument of J_n. The expression for E_z in the cladding region can be obtained immediately from Eq. 10.31 by replacing ϵ_2 with ϵ_1, u with w, A with B, and J_n or J_n' with K_n and K_n', respectively. Similar calculations for H_z yield for $r \le a$

$$H_z = \frac{jA}{\omega \mu} \left[\frac{u}{a} J_n'\left(\frac{u}{a} r\right) \binom{\cos n\phi}{\sin n\phi} \cos \phi - \frac{n}{r} J_n\left(\frac{u}{a} r\right) \binom{-\sin n\phi}{\cos n\phi} \sin \phi \right] \tag{10.32}$$

For $r > a$, H_z is obtained by replacing A with B, u with w, and J_n and J_n' with K_n and K_n', respectively.

10.4.2 Boundary Conditions and the Guidance Condition

To determine the constants A and B and obtain the guidance condition for the propagation constant k_z it is necessary to apply the boundary conditions at the interface between the core and cladding. This requires that the tangential components of **E** and **H** be continuous at $r = a$. Note that since **H** is not independent of **E** under the assumed form of solutions, it is sufficient to simply match tangential components of **E**, E_ϕ and E_z; the tangential components of **H** will therefore automatically satisfy the boundary conditions. The ϕ component of E_y is simply given by $E_\phi = E_y \cos \phi$. Thus from Eqs. 10.17, 10.25, and 10.26, we have after matching E_ϕ across the boundary at $r = a$,

$$AJ_n(u) = BK_n(w) \tag{10.33}$$

Equating E_z at $r = a$ and using Eq. 10.33 to relate A to B yields under the approximation $\epsilon_2 \cong \epsilon_1$

$$uK_n(w)J'_n(u) = wJ_n(u)K'_n(w) \tag{10.34}$$

which is the guidance condition. The above relation can be somewhat simplified by making use of the recurrence relations for Bessel functions:

$$J'_n(z) = -(n/z)J_n(z) + J_{n-1}(z) \tag{10.35}$$

$$K'_n(z) = -(n/z)K_n(z) - K_{n-1}(z) \tag{10.36}$$

Substituting Eqs. 10.35 and 10.36 into Eq. 10.34 yields straightforwardly

$$u\frac{J_{n-1}(u)}{J_n(u)} = -w\frac{K_{n-1}(w)}{K_n(w)} \tag{10.37}$$

Trancendental equation 10.37 is a function of the propagation constant k_z, the waveguide material parameters, and the integer n. For any choice of n there exists, in general, m solutions for k_z which we designate by $(k_z)_{nm}$ and the associated linearly polarized mode by the notation LP_{nm}. Note that Eq. 10.37 is valid whether Ψ was initially chosen to be proportional to $\cos n\phi$ or $\sin n\phi$. Further, the electric field could have initially been assumed polarized along $\hat{x}$ rather than along $\hat{y}$. This implies that, in general, there are four solutions for each LP_{nm} mode. These modes are said to be fourfold degenerate. It is important to note that this degeneracy is the result of the assumption that the fractional difference in refractive index between core and cladding is very small. An exact solution to Maxwell's equations would show that in fact when this is not the case the modes separate from each other and the concept of the LP-mode designation loses sense. There is one exception to the fourfold degeneracy, however. We see from Eq. 10.26 that when $n = 0$ only one solution for **E** exists for each polarization. Thus the LP_{0m} mode is only twofold degenerate.

The value of $(k_z)_{nm}$ must, in general, be obtained from Eq. 10.37 by numerical computation. In analogy with the asymmetric slab dielectric waveguide let us define the parameter

$$b = \frac{\epsilon_{\text{eff}} - \epsilon_1}{\epsilon_2 - \epsilon_1}$$

where

$$\frac{\omega}{k_z} = \frac{1}{\sqrt{\mu \epsilon_{\text{eff}}}}$$

Using the usual definition for normalized waveguide thickness v, a plot of b versus v can be computed from Eq. 10.37 for the various LP modes and is shown in Figure 10.11 for a number of solutions.

10.4.3 Cutoff

The cutoff condition is readily obtained from Eq. 10.37. At cutoff, $w \to 0$ and $u \to v$ so that we obtain a constraint on the value of v at cutoff given by

$$J_{n-1}(v) = 0 \qquad (10.38)$$

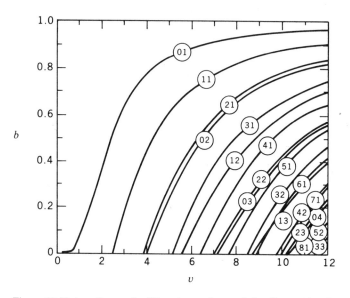

Figure 10.11 b–v diagram for LP modes on the step-index fiber. (After Gloge (Ref. 7). Reproduced by permission of the author and the Optical Society of America.)

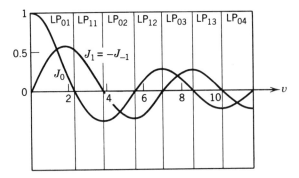

Figure 10.12 Solution to the cutoff condition, Eq. 10.38. (After Gloge (Ref. 7). Reproduced by permission of the author and the Optical Society of America.)

In general, there will be m solutions for v for each value of n, denoted by v_{nm}. Thus, for example, the cutoff values of v for the LP_{0m} modes are given by the m zeroes of the Bessel function $J_{-1}(v) = J_1(v)$. Similarly, the cutoff values for the LP_{1m} modes are obtained from the zeroes of $J_0(v)$ as shown in Figure 10.12. The relative ordering of v at cutoff for the various modes can be obtained by examining the ordering of the zeroes of their respective Bessel functions. We see, for example, that the LP_{01} mode has a cutoff value of $v = 0$, which implies it propagates at all frequencies. The next higher order is the LP_{11} which starts to propagate for $v > 2.405$. Because of its widespread use we mention here that the lowest order mode is often referred to as the HE_{11} mode, a notation that is conventionally used when an exact solution to Maxwell's equation is obtained.

Example 10.2

A weakly guiding multimode fiber has a core refractive index $n_2 = 1.46$, cladding index $n_1 = 1.45$, and radius $a = 5$ μm. It is excited by a light-emitting diode (LED) having $\lambda_0 = 0.90$ μm.

(a) How many LP modes will the fiber support?
(b) What is the effective index for the LP_{02} mode?

Solution

(a)

$$v \cong \frac{2\pi a}{\lambda_0} \sqrt{2\,\Delta n\, n}$$

$$= 2\pi \left(\frac{5}{0.9}\right) \sqrt{2\,(0.01)\,1.46}$$

$$= 5.96$$

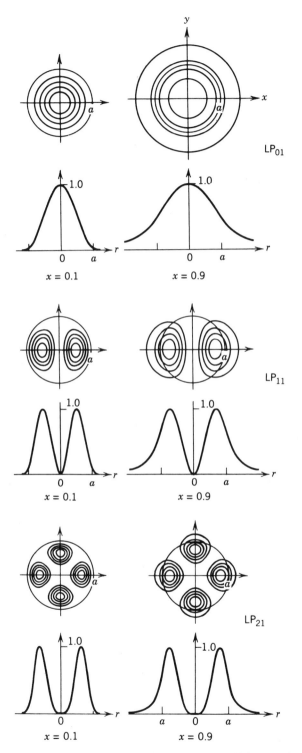

Figure 10.13 Normalized intensity plots for several LP modes for small and large values of $x = u^2/v^2$. (After Okashi (Ref. 14). Reproduced by permission of the author.)

From Figure 10.11, the LP_{01}, LP_{11}, LP_{21}, LP_{02}, LP_{31}, and LP_{12} modes will propagate. All are fourfold degenerate except for LP_{01} which is twofold degenerate. Therefore, 22 modes will propagate.

(b) From Figure 10.11, for $v = 5.96$, $b \cong 0.42$ for the LP_{02} mode. Thus,

$$n_{\text{eff}} = \sqrt{(n_2^2 - n_1^2)b + n_1^2} = 1.454$$

10.4.4   Power Density Distribution

It is of interest to investigate the spatial variation of the intensity of the LP modes at any arbitrary fiber cross section, $z = $ constant. Physically, if we were to view this intensity with the eye, we would measure the $\hat{z}$ component of the time-average Poynting power. From Eq. 10.17 this is given by

$$\langle \mathbf{S}(r, \phi) \rangle \cdot \hat{z} = \frac{1}{2\eta} |\Psi(r, \phi)|^2$$

Substituting for Ψ from Eqs. 10.25 and 10.26 and making use of Eq. 10.33 to relate the field amplitudes in the core and cladding regions gives

$$\langle \mathbf{S}(r, \phi) \rangle \cdot \hat{z} = \frac{|A|^2}{2\eta} \begin{pmatrix} \cos^2 n\phi \\ \sin^2 n\phi \end{pmatrix} \begin{cases} J_n^2 (ur/a) & r \leq a \\ \dfrac{J_n^2(u)}{K_n^2(w)} K_n^2 (wr/a) & r > a \end{cases}$$

Normalized intensity plots using an assumed $\cos^2 v\phi$ variation for the LP_{01}, LP_{11}, and LP_{21} mode are shown in Figure 10.13 for relatively small and large values of the quantity $x \equiv u^2/v^2$. For x small, the mode is relatively tightly bound whereas for x approaching unity the mode approaches cutoff.

10.5   LOSS MECHANISMS IN OPTICAL FIBERS

Attenuation of signals propagating along an optical fiber is of particular importance because it determines the maximum allowed spacing between repeater stations. Attenuation in fibers is the result of three principal mechanisms: (1) absorption by the fiber material, (2) scattering by imperfections and intrinsic density fluctuations in the fiber material, and (3) losses generated by energy radiated out of the guide as the result of fiber bends and geometrical nonuniformities.

10.5.1   Absorption

Due to significant strides in the ability to remove OH radical contamination during the fiber fabrication process, absorption levels in silica fibers are presently determined by infrared absorption associated with the inherent chemical bonding of the constituent atoms of the silica or doped silica fiber. This absorption is wavelength dependent, having a peak in the 9-μm range. As shown in Figure 10.14*a,* the tail end of the absorption curve determines the useful long wavelength transmission limit for solid state optical sources which typically radiate wavelengths in the 0.8- to 1.6-μm region.

10.5.2 Scattering

The second loss mechanism inherent in the fiber is the result of Rayleigh scattering. This scattering is caused by random spatial fluctuations in density which are frozen into the glass material upon solidification. These density fluctuations result in corresponding random spatial variations in refractive index. The attenuation coefficient for such scattering is proportional λ^{-4} and is shown in Figure 10.14*a* for a typical silica single-mode fiber. From the figure, it is clear that the inherent loss mechanisms, absorption and scattering, define a low-loss optical "window" which determines a range of usable source wavelengths for optical communication centered around the 1.3- to 1.6-μm region. As shown in Figure 10.14*b,* for low OH content fibers, losses of less than 0.2 dB/km are obtainable.

10.5.3 Radiation Bending Losses

The attenuation of guided waves propagating along a section of curved waveguide was treated in Section 5.4 for the slab or rectangular waveguide geometry. It was shown

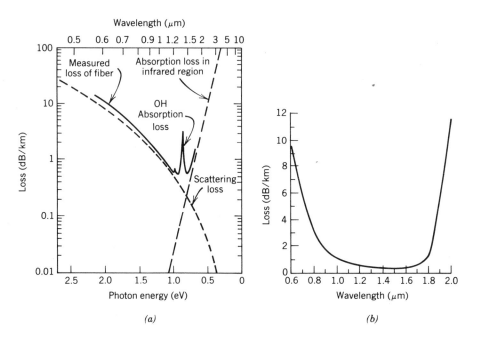

Figure 10.14 (*a*) Various fiber material contributions to total propagation loss (After Moriyama *et al.* (Ref. 13).) (*b*) Low OH content fiber. (After Osanai *et al.* (Ref. 15).) Reproduced by permission of the Institution of Electrical Engineers.)

that the attenuation coefficient α depended exponentially on the bending radius R. A detailed analysis of bending losses for the fiber guide shows that the attenuation for such fibers is also dependent on radius of curvature in the same exponential fashion as for the slab waveguide. In Ref. (17) the normalized attenuation coefficient αa for the fundamental mode on the step-index fiber is calculated and is shown to be given to good approximation by

$$\alpha a = \frac{1}{(\pi w R/a)^{1/2}} \left(\frac{u}{v}\right)^2 \exp\left[\frac{-2}{3}\left(\frac{w^3}{v^2}\right)\left(2\frac{\Delta n}{n}\right)\frac{R}{a}\right] \qquad (10.39)$$

where R is the waveguide bending radius and all other parameters have been defined in Sections 10.3 and 10.4. The attenuation losses are particularly significant for monomode fibers because, as was previously shown, they can exhibit material propagation losses of only a few tenths of a decibel per kilometer. Thus it is necessary in stringing such a fiber to keep radiation losses significantly below this level. Figure 10.15 shows the typical dependence of the attenuation of the fundamental mode on a step-index fiber calculated from expression 10.39 as a function of the fractional index difference between core and cladding, $\Delta n/n$, for several bending radii.

Example 10.3

Compute the power loss in decibels for a monomode step-index fiber having a bending radius of 2 cm and the following other waveguide parameters

$$n_2 = 1.45 \qquad \lambda_0 = 1.25 \ \mu m$$
$$\frac{\Delta n}{n} = 0.3\% \qquad a = 4.25 \ \mu m$$

Solution

From Eq. 10.13, the propagation constant for a monomode step-index fiber was shown to be given approximately by

$$k_z^2 = k_2^2 \left[1 - \left(\frac{\Delta}{n_2}\right)^2 \frac{1}{v^2}(1 + \ln v^2)\right]$$

The parameter u^2 is related to k_z by

$$u^2 = (k_2^2 - k_z^2) a^2$$
$$= \left(\frac{\Delta}{n_2}\right)^2 \left(\frac{k_2 a}{v}\right)^2 (1 + 2 \ln v)$$
$$= 1 + 2 \ln v$$

Similarly,

$$w^2 = (k_z^2 - k_1^2) a^2 = v^2 - (1 + 2 \ln v)$$

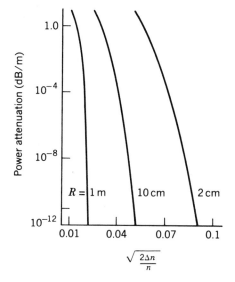

Figure 10.15 Radiation bending loss for the fundamental mode of a step-index fiber: $u = 1.527$, $v = 2.2$, $w = 1.584$, $\lambda_0 = 1$ μm, $n = 1.5$. (After Snyder *et al.* (Ref. 17). Reproduced by permission of the Institution of Electrical Engineers.)

The quantity $v = k_0 a \Delta \cong (k_2 a)\sqrt{2\,\Delta n/n}$ can therefore be computed from the waveguide parameters as

$$v = \frac{2\pi}{\lambda_0} n_2 a \sqrt{2\,\Delta n/n} = 2.400$$

The corresponding values for u and w are

$$u = 1.659$$

$$w = 1.735$$

Substituting the values for u, v, w and the waveguide parameters into Eq. 10.39 yields for the attenuation coefficient

$$\alpha = 2.7 \times 10^{-5} \text{ m}^{-1}$$

The power attenuation in decibels after 1 km is therefore

$$(P_t)_{\text{dB}} = -10 \log[e^{-\alpha(1000)}] = 0.12 \text{ dB/km}$$

10.6 SIGNAL DISTORTION DUE TO DISPERSION

A very important consideration for optical fibers is their information capacity. For digital communication systems, the transmitted data usually consist of a train of binary pulses coded in some fashion. The information capacity of such a system is maximized by sending as many pulses per unit time along the fiber as possible, which implies minimizing pulse duration.

One of the most significant limitations to obtainable data rates is the result of the nonlinear relationship that generally exists between frequency and propagation constant, known as dispersion. In Chapter 2, dispersion was shown in an elementary fashion to result in a pulse envelope or group velocity different from the carrier or phase velocity. We show in this section that dispersive systems also cause data pulses to continually broaden as they propagate. Thus as the separation between transmitter and receiver is increased, the maximum allowable data rate must be appropriately reduced to prevent adjacent pulses from overlapping as they broaden. The dispersion is shown to be influenced by three primary factors: (1) material dispersion resulting from the frequency-dependent refractive index of the fiber material itself, (2) waveguide dispersion due to the fact that the propagation constant of every mode depends on the waveguide dimensions relative to a wavelength, and (3) intermodal dispersion resulting from distribution of pulse energy over a number of modes, each having a propagation constant with a different frequency dependence.

10.6.1 Effect of a Dispersive System on a Gaussian Pulse

Let us examine the effect of transmitting along a dispersive fiber a pulse having a Gaussian distribution in time. This particular waveform is chosen for mathematical convenience but the results obtained are applicable to pulses of arbitrary shape. Figure 10.16 shows schematically how the pulse might be generated. A monochromatic optical beam at ω_0 is passed through a modulator producing a Gaussian-shaped pulse in the time domain which is subsequently focused upon the input end of the fiber. For simplicity, we assume that only one mode of the fiber is excited by the incident radiation. The time dependence of the electromagnetic fields associated with the mode at the input end $z = 0$ is therefore of the form (real part implied)

$$f(t, z = 0) = e^{-(1/2)(\sigma t)^2} e^{j\omega_0 t} \tag{10.40}$$

where $\Delta t = 1/\sigma$ is the pulse half-width.

The spectrum $F(\omega)$ of $f(t)$ can be obtained from the definition of the Fourier transform,

$$F(\omega) = \int_{-\infty}^{\infty} dt\, f(t) e^{-j\omega t} \tag{10.41}$$

The corresponding inverse Fourier transform is then given by

$$f(t) = \frac{1}{2\pi} \int_{-\infty}^{\infty} d\omega\, F(\omega) e^{j\omega t} \tag{10.42}$$

Using the Fourier transform pair definition above, the spectrum of $f(t, z = 0)$ is found

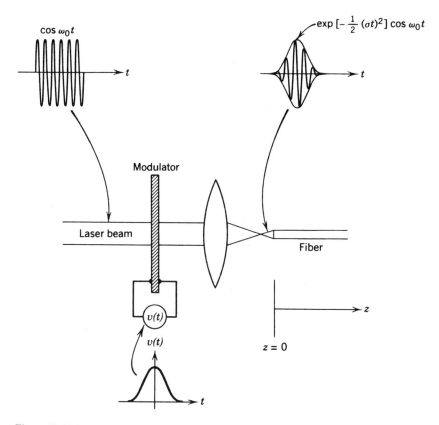

Figure 10.16 Modulated light pulse coupled into an optical fiber.

also to be Gaussian and given by

$$F(\omega, \sigma) = \frac{\sqrt{2\pi}}{\sigma} \exp\left[-\frac{1}{2}\left(\frac{\omega - \omega_0}{\sigma}\right)^2 \right] \tag{10.43}$$

The dependence of the Fourier transform on the pulse width has been indicated explicitly by including it in the argument of F. This will prove convenient in the derivation that follows. The time- and frequency-domain representations of the pulse at the input of the fiber are shown in Figure 10.17. Note that as is expected from the properties of the Fourier transform, the pulse width in the frequency domain is the reciprocal of its duration in the time domain.

Now consider the shape of the input pulse at some arbitrary observation point z along the line. To analyze this problem, it is only necessary to realize that the input pulse may be regarded as a superposition of source signals with different frequencies. The amplitude of each of these sources is given by the Fourier spectrum $F(\omega, \sigma)$ shown in Figure 10.17. Each Fourier component will have associated with it a propagation

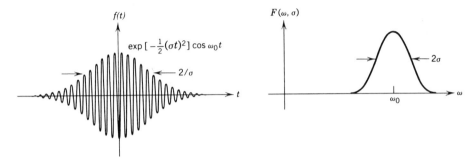

Figure 10.17 Time dependence and Fourier spectrum of optical signal pulse at input to fiber.

constant $k_z(\omega)$ and will therefore be phase shifted at any arbitrary output plane z relative to its value at the input by an amount $\Delta\phi = -k_z(\omega)z$.

Thus in absence of loss, the optical fiber can be viewed as a simple linear system having a transfer function

$$H(\omega) = e^{-jk_z(\omega)z}$$

In the frequency domain, the input to the system is $F(\omega, \sigma)$ so that the output $Y(\omega)$ is

$$Y(\omega) = H(\omega)F(\omega, \sigma)$$

Taking the inverse Fourier transform of Y, the output will therefore have a time dependence given by

$$f(t, z) = \frac{1}{2\pi} \int_{-\infty}^{\infty} d\omega \, F(\omega, \sigma)e^{-jk_z(\omega)z}e^{j\omega t} \tag{10.44}$$

In theory, for any given $k_z(\omega)$, Eq. 10.44 can be solved numerically. In practice, however, several valid simplifying approximations can be made allowing the solution to be expressed in closed form.

With reference to Figure 10.17, it is observed that $F(\omega, \sigma)$ is small outside the frequency range $|\omega - \omega_0| > \sigma$. Since the integrand in Eq. 10.44 is approximately zero outside this frequency range, $k_z(\omega)$ may therefore be expanded in a Taylor series about ω_0 and the first few terms in the series kept without significant loss in accuracy. That is,

$$k_z(\omega) \cong k^{(0)} + \Delta\omega \, k^{(1)} + \frac{\Delta\omega^2}{2} k^{(2)}$$

where

$$\Delta\omega \equiv \omega - \omega_0$$

and

$$k^{(n)} \equiv \left. \frac{\partial^n k_z}{\partial \omega^n} \right|_{\omega = \omega_0}$$

Note that for larger excursions in $\Delta \omega$ where the higher order terms in the series expansion of k_z might become large, $F(\omega, \sigma)$ is small (approximately zero). Thus $f(t, z)$ is given to good approximation by

$$f(t, z) \cong \frac{1}{2\pi} \int_{-\infty}^{\infty} d\omega \, F(\omega, \sigma) \exp\left[j\left(\omega t - k^{(0)}z - \Delta\omega \, k^{(1)}z - \frac{\Delta\omega^2}{2} k^{(2)}z \right) \right] \quad (10.45)$$

From the definition of $F(\omega, \sigma)$ given in Eq. 10.43, however, we observe that the term in the integrand

$$F(\omega, \sigma) \, e^{-j(\Delta\omega^2/2)K^{(2)}z}$$

can be written as

$$\frac{\tilde{\sigma}}{\sigma} F(\omega, \tilde{\sigma})$$

where the complex "width" $\tilde{\sigma}$ is defined by the relation

$$\left(\frac{1}{\tilde{\sigma}} \right)^2 = \frac{1}{\sigma^2} + jk^{(2)}z \quad (10.46)$$

Therefore Eq. 10.44 becomes

$$f(\tilde{t}, z) = \frac{\tilde{\sigma}}{\sigma} e^{-jk^{(0)}z} e^{+j\omega_0 k^{(1)}z} \left[\frac{1}{2\pi} \int_{-\infty}^{\infty} d\omega \, F(\omega, \tilde{\sigma}) \exp(j\omega \tilde{t}) \right]$$

where the new time variable $\tilde{t}$ has been defined by

$$\tilde{t} \equiv t - k^{(1)}z$$

The output power will be proportional to the square of the magnitude of $f(t, z)$ or

$$P_{\text{out}} \propto \left| \frac{\tilde{\sigma}}{\sigma} \right|^2 \left| \frac{1}{2\pi} \int_{-\infty}^{\infty} d\omega \, F(\omega, \tilde{\sigma}) \exp(j\omega \tilde{t}) \right|^2$$

The integral above is seen to be simply the inverse Fourier transform of a Gaussian pulse with spectral "width" $\tilde{\sigma}$, but with a new time axis $\tilde{t}$. Comparing the Fourier transform pair in Eqs. 10.40 and 10.43 we see that the power in the output pulse will be

$$P_{out} = \left| \frac{\tilde{\sigma}}{\sigma} \right|^2 \left| \exp\left(-\frac{\tilde{\sigma}^2 \tilde{t}^2}{2} \right) \right|^2$$

Noting that since

$$\tilde{\sigma}^2 = \text{Re}(\tilde{\sigma}^2) + j \, \text{Im}(\tilde{\sigma}^2)$$

then

$$\left| \exp\left(-\frac{\tilde{\sigma}^2 \tilde{t}^2}{2} \right) \right|^2 = |\exp[-(1/2)\text{Re}(\tilde{\sigma}^2)\tilde{t}^2]|^2 \, |\exp[-(j/2) \, \text{Im}(\tilde{\sigma}^2)\tilde{t}^2]|^2$$

$$= \exp[-\text{Re}(\tilde{\sigma}^2)\tilde{t}^2]$$

so that

$$P_{out} = \frac{|\tilde{\sigma}|^2}{\sigma^2} \exp[-\text{Re}(\tilde{\sigma}^2)\tilde{t}^2]$$

Using the definition of $\tilde{\sigma}^2$ from Eq. 10.46 we obtain

$$P_{out} = \frac{1}{[1 + (\sigma^2 k^{(2)}z)^2]^{1/2}} \exp\left(-\frac{\sigma^2}{1 + (\sigma^2 k^{(2)}z)^2} \, \tilde{t}^2 \right) \tag{10.47}$$

Note that the peak amplitude of the output pulse power occurs for $\tilde{t} = 0$. Thus to remain at the coordinate z of peak power for all time requires that

$$\tilde{t} = t - k^{(1)}z = 0$$

This implies that the rate of change of the observation point z with time, that is the observer velocity, is given by

$$\frac{dz}{dt} = \left. \frac{\partial \omega}{\partial k_z} \right|_{\omega = \omega_0} = v_g$$

in agreement with the simplified analysis given in Chapter 2.

Additionally we see from Eq. 10.47 that the new pulse duration is given by

$$\Delta \tilde{t} = \left[\frac{1}{\sigma^2} + (\sigma k^{(2)}z)^2 \right]^{1/2} \tag{10.48}$$

and the peak amplitude of the pulse power has been reduced by the factor

$$\frac{P_{peak}(z)}{P_{peak(z=0)}} = \frac{1}{[1 + (\sigma^2 k^{(2)} z)^2]^{1/2}} \tag{10.49}$$

Note that pulse spreading is a function of three factors: (1) bandwidth of the input pulse, (2) dispersion of the waveguide through the factor $k^{(2)}$, and (3) distance the pulse travels. A simple physical explanation can be given for these results. For a dispersive waveguide, the group velocity v_g is a function of ω. The delay time τ required for a monochromatic signal to travel a distance z along the fiber is therefore

$$\tau(\omega) = \frac{z}{v_g(\omega)}$$

We observe that the travel time is frequency dependent. Now suppose that the signal is not monochromatic but rather is distributed uniformly over a bandwidth $\Delta\omega$. The corresponding spread in arrival times $\Delta\tau$ will therefore be

$$\Delta\tau = z \frac{d(1/v_g)}{d\omega} \Delta\omega$$

But

$$\frac{1}{v_g} = \frac{dk_z}{d\omega}$$

so that

$$\Delta\tau = zk^{(2)} \Delta\omega$$

Identifying $\Delta\omega$ with the bandwidth parameter σ in the previous derivation we see that

$$\Delta\tau = zk^{(2)} \sigma \tag{10.50}$$

Comparing expression 10.50 with Eq. 10.48 shows that when the pulse spread is much larger than the initial pulse width the two results are identical.

10.6.2 Material Dispersion

Material dispersion results from the fact that the refractive index of optical materials is a function of wavelength. This phenomenon of course is what is responsible for refraction of white light by a prism.

Let us consider a plane wave propagating through a uniform medium having refractive index $n(\lambda)$. The propagation constant k for this wave is given by

$$k = n\frac{\omega}{c}$$

We wish to compute the parameter $k^{(2)}$ since from Eq. 10.50 this term characterizes the effect of the material on pulse broadening. Taking the derivative of k with respect to ω yields

$$\frac{dk}{d\omega} = \frac{1}{c}\left(n + \omega\frac{dn}{d\omega}\right)$$

Since the variation in refractive index is generally described in terms of wavelength, we express the dependence of n on λ_0 rather than ω, where λ_0 is the free space wavelength defined by

$$f\lambda_0 = c \tag{10.51}$$

Now

$$\frac{dn}{d\omega} = \frac{dn}{d\lambda_0}\frac{d\lambda_0}{d\omega} = \frac{dn}{d\lambda_0}\left(\frac{-2\pi c}{\omega^2}\right) \tag{10.52}$$

therefore

$$\frac{dk}{d\omega} = \frac{1}{c}\left(n - \lambda_0\frac{dn}{d\lambda_0}\right)$$

Taking a second derivative with respect to ω yields

$$k^{(2)}\sigma = \frac{d^2k}{d\omega^2}\Delta\omega = \frac{d}{d\lambda_0}\left(\frac{dk}{d\omega}\right)\Delta\lambda_0$$

$$= \frac{1}{c}\left(-\lambda_0\frac{d^2n}{d\lambda_0^2}\right)\Delta\lambda_0$$

Thus, the pulse spread $\Delta\tau_{mat}$ due to material dispersion is

$$\Delta\tau_{mat} = k^{(2)}z\sigma = -z\left(\frac{\lambda_0}{c}\frac{d^2n}{d\lambda_0^2}\right)\Delta\lambda_0 \tag{10.53}$$

Figure 10.18 shows a plot of the term in parenthesis for pure silica. Note that the pulse broadening resulting from material dispersion goes to zero in pure silica for $\lambda_0 \cong 1.3$

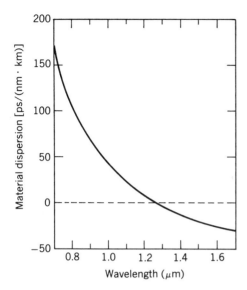

Figure 10.18 Material dispersion as a function of wavelength for a fiber made of pure silica. (After Fleming (Ref. 4). Reproduced by permission of the Institution of Electrical Engineers.)

μm. Further, from Eq. 10.53 it is observed that the amount of pulse broadening is directly proportional to the spectral width of the optical source.

Example 10.4

An LED having $\lambda_0 = 0.9$ μm and $\Delta\lambda_0 = 0.025$ μm excites a SiO_2 optical fiber. What is the material dispersion over 1 km?

Solution

From Figure 10.18, at $\lambda_0 = 0.9$ μm the dispersion is 70×10^6 ps/m^2. Thus

$$\Delta\tau_{mat} = \frac{70 \times 10^6 \text{ ps}}{m^2} \times 1 \times 10^3 \text{ m} \times 0.025 \times 10^{-6} \text{ m}$$

$$= 1750 \text{ ps}$$

10.6.3 Waveguide Dispersion

Because a waveguide is an inherently dispersive system, it is of interest to determine its effect on pulse distortion. The influence of dispersion is of particular importance for monomode fibers since as discussed previously these are the prime candidates for long-distance, high data rate communication systems. If we operate our monomode fiber at a wavelength near where material dispersion is small or zero then as shown previously in section 10.6.1, the pulse spread $\Delta\tau_{wg}$, is given in terms of the waveguide group velocity v_g by

$$\Delta\tau_{wg} = z \frac{dv_g^{-1}}{d\omega}\left(\frac{-\omega^2}{2\pi c}\right)\Delta\lambda_0$$

Since $v_g^{-1} = dk_z/d\omega = (dk_z/dv)(dv/d\omega)$, the approximate expression for k_z given by Eq. 10.14 can be used along with the definition for v in Eq. 10.12. We obtain for v_g^{-1}

$$v_g^{-1} = \frac{n_2}{c}\left[1 - \left(\frac{\Delta n}{n}\right)\frac{1}{v^2}(1 - \ln v^2)\right]$$

Differentiating a second time with respect to ω yields straightforwardly for $\Delta\tau_{wg}$

$$\Delta\tau_{wg} = \frac{-z}{c\lambda_0}\Delta n\, D\,\Delta\lambda_0 \qquad (10.54)$$

where the distortion parameter D is defined by

$$D = 4(1 - \ln v)/v^2$$

and is plotted in Figure 10.19 along with an exact calculation for the fundamental mode (Ref. 6).

It is observed from Eq. 10.54 that for a given normalized frequency v, the pulse spread is proportional to Δn. Thus weakly guiding fibers exhibit small distortion. Additionally from Figure 10.19 it is observed that for a specific value of v, pulse

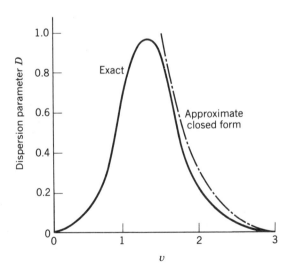

Figure 10.19 Comparison of approximate and exact expression for the distortion parameter D. (After Gloge (Ref. 6). Reproduced by permission of the author and the Optical Society of America.)

spreading resulting from waveguide dispersion goes to zero. This result is significant because if waveguide dispersion can be made equal to zero at the same wavelength for which material dispersion disappears then at this wavelength, distortion-free pulses can be sent over large distances. Alternatively, if the waveguide and material dispersions can be made equal but of opposite sign the net dispersion can be reduced to zero. This issue is addressed in the problems. Further, by use of more exotic refractive-index profiles such as those shown in Figure 10.20 the wavelength where total dispersion goes to zero can be shifted or in the case of the segmented waveguide, extended over a large range of wavelengths. Compensation over a wide range of λ_0 is of particular significance in increasing fiber information capacity; it allows signals from several different laser sources tuned to different wavelengths to be transmitted simultaneously on the same fiber, a technique known as wavelength multiplexing.

Example 10.5

A single-mode step-index fiber has a core radius $a = 2$ μm, refractive index $n_2 = 1.461$, cladding index $n_1 = 1.456$, and is excited by a multimode laser having $\lambda_0 = 0.85$ μm and $\Delta\lambda_0 = 1 \times 10^{-3}$ μm. What is the pulse spread $\Delta\tau_{wg}$ due to waveguide dispersion over a kilometer?

Solution

$$\Delta\tau_{wg} = \frac{-z}{c\lambda_0} \Delta n \, D \, \Delta\lambda_0$$

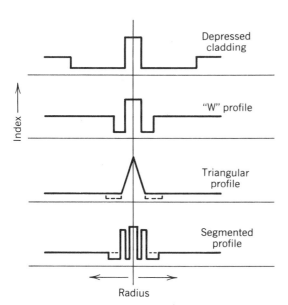

Figure 10.20 Refractive-index profiles of a number of weakly guiding fibers. (After Keck (Ref. 9). Copyright © 1983, IEEE. Reproduced by permission.)

where

$$D = 4(1 - \ln v)/v^2$$

Now

$$v \cong \frac{2\pi a}{\lambda_0} \sqrt{2 \, \Delta n \, n} = 2\pi \left(\frac{2}{0.85}\right) \sqrt{2 \, (0.005) \, 1.461}$$

$$= 1.787$$

Therefore,

$$D = 4(1 - \ln 1.787)/1.787^2 = 0.525$$

and

$$|\Delta\tau_{wg}| = \frac{1 \times 10^3}{3 \times 10^8 \times 8.5 \times 10^{-7}} (0.005) (0.525) (1 \times 10^{-9}) = 10.3 \text{ ps}$$

10.6.4 Intermodal Dispersion

The last factor responsible for signal distortion is intermodal dispersion which results from the fact that at any given frequency, each waveguide mode propagates at a different group velocity.

This factor is of course eliminated in monomode fibers. Although a rigorous analysis of such distortion is rather complicated, a good estimate of the size of the effect can be obtained from simple geometric optics arguments. With reference to Figure 10.21, the propagation angle for rays inside a planar dielectric guide ranges from the critical angle θ_c for modes near cutoff (highest order modes) to $\pi/2$ for modes far from cutoff (lowest order modes). Thus modes near cutoff have velocities measured along the waveguide axis which are less than those modes which are far from cutoff. For a fiber of length L the difference in arrival time $\Delta\tau_{int}$ for the two limiting situations is

$$\theta_{min} \cong \frac{\pi}{2} - \sqrt{2\Delta n/n}$$

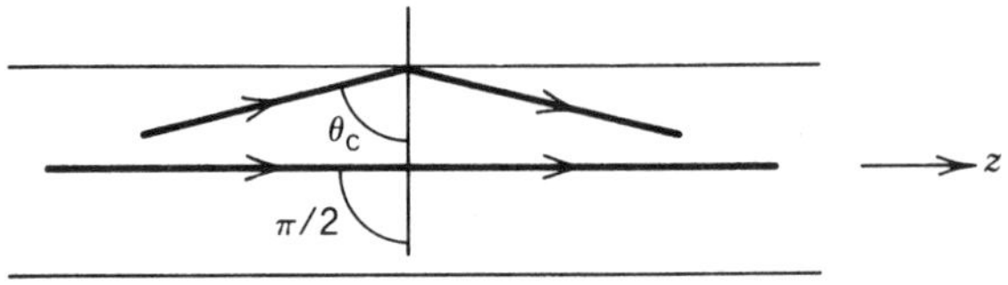

Figure 10.21 Propagation angle for rays inside a planar dielectric waveguide.

From Figure 10.20 we see that

$$v_{max} = \omega/k_2$$

$$v_{min} = (\omega/k_2) \sin \theta_c = (\omega/k_2)(n_1/n_2)$$

so that we obtain as an estimate for the intermodal pulse spread

$$\Delta\tau_{int} \cong \frac{L}{c} \Delta n$$

It is important to observe that intermodal dispersion is independent of the spread in wavelength of the source whereas waveguide and material dispersion are not.

It is also of interest to compute approximately the intermodal dispersion for a multimode parabolic graded-index fiber. As with the step-index fiber, we use the planar graded-index waveguide model discussed in Chapter 5 to facilitate the analysis. The propagation constant for the pth mode was shown to be (Eq. 5.13)

$$(k_z)_p \cong k_0 n_{max} [1 - (2p + 1)/(k_0 x_0 n_{max})]^{1/2}$$

where $2\pi x_0$ is the periodicity of the rays in the waveguide along the direction of propagation. Unlike step-index guides, the periodicity for all paraxial modes is the same. Thus, we see that provided the second term in brackets above is small so that the binomial expansion can be used:

$$(1 - x)^{1/2} \cong 1 - \tfrac{1}{2}x$$

then the group velocity for all modes is equal to n_{max}/c and no waveguide dispersion exists. Physically this occurs because as shown in Figure 10.22, although higher order

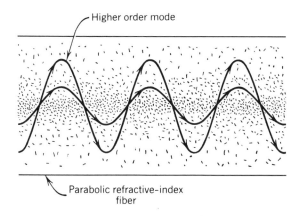

Figure 10.22 Ray paths for two different modes in a parabolic refractive-index fiber.

modes travel a greater physical distance within the guide, a larger percentage of their path is through a proportionately lower refractive index and thus higher velocity material. For the parabolic index variation, the two effects nearly exactly compensate each other yielding to first order the same propagation velocity measured along the waveguide axis for all modes.

Example 10.6

What is the intermodal dispersion $\Delta\tau_{int}$ for a 1-km-long step fiber having $\Delta n = 0.005$?

Solution

$$\Delta\tau_{int} = \frac{L}{c}\,\Delta n = \frac{1 \times 10^3}{3 \times 10^8}\,(0.005) = 1.67 \times 10^4 \text{ ps}$$

Note that the value of intermodal dispersion is much greater than $\Delta\tau_{wg}$ or $\Delta\tau_{max}$ computed in the previous example.

PROBLEMS

10.1 (a) Show that the minimum bounce angle θ_{min} for an asymmetric slab dielectric waveguide having core refractive index n_2 and substrate index n_3 is given approximately by

$$\theta_{min} \cong \frac{\pi}{2} - \sqrt{2\Delta n/n}$$

where $\Delta n = n_2 - n_3$ and $n_2 \cong n_3 \equiv n$.
(b) Compute θ_{min} for a waveguide having $n = 1.46$ and $\Delta n = 0.0025$.

10.2 A step-index optical fiber, shown in Figure 10.23, is to be excited by illuminating it with a light source placed at one end. Because the source has a finite aperture, the emitted radiation will spread out as shown. Using ray arguments, the fiber can only accept those incident rays which enter at an angle such that they are totally internally reflected at the core–cladding interface. The fiber numerical aperture NA, which is a measure of the acceptance angle, is defined as

$$NA = \sin\theta_i$$

(a) Show that if n_2 and n_1 are the fiber core and cladding indices,

$$NA = \sqrt{n_2^2 - n_1^2}$$

(b) What is the numerical aperture and θ_i for a fiber having $n_2 = 1.46$ and $n_1 = n_2 - \Delta n$, where $\Delta n = 0.004$?

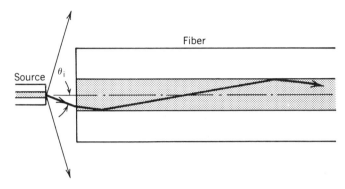

Figure 10.23 Excitation of a step-index optical fiber by a light source.

10.3 A weakly guiding fiber has a Gaussian variation in refractive index given by

$$n^2(r) = n_1^2 + (n_2^2 - n_1^2)e^{-(r/a)^2}$$

(a) Assuming a solution $\psi(r)$ having the functional form given in Eq. 10.9, show using the variational technique of Section 10.3.1 that the propagation constant k_z is given by

$$k_z^2 = k_2^2 - \frac{1}{r_0^2} - \frac{v^2}{a^2}\left(\frac{r_0^2}{a^2 + r_0^2}\right)$$

(b) Show that the best choice for r_0 is

$$r_0 = \frac{a}{(v - 1)^{1/2}}$$

10.4 A monomode step-index fiber has the following specifications:

$$\lambda_0 = 1.3 \ \mu\text{m} \qquad n_2 = 1.46$$
$$a = 2\lambda_0 \qquad \Delta n = 0.005$$

Compute the phase and group velocities.

10.5 Make a plot of the normalized power density S_n versus r/a for the monomode waveguide described in problem 10.4.

10.6 A weakly guiding multimode waveguide has the following specifications:

$$\lambda_0 = 1.0 \ \mu\text{m} \qquad n_2 = 1.5$$
$$a = 5.5 \ \mu\text{m} \qquad \Delta n = 0.012$$

How many LP modes can propagate on the fiber?

10.7 (a) Determine the maximum radius a that a weakly guiding fiber may have for it to support only the fundamental mode. Assume $\lambda_0 = 0.63$ μm, $n_2 = 1.42$, and $\Delta n = 0.0025$.

(b) For the calculated radius above, over what frequency range can the source be varied without exciting higher order modes?

10.8 What is the maximum bending radius R that a weakly guiding fiber may have in order that the radiation bending losses for the fundamental mode are less than 1 dB/km? The fiber has the following specifications:

$$n_2 = 1.43 \qquad \lambda_0 = 1.3 \text{ μm}$$
$$\Delta n = 0.004 \qquad a = 3.9 \text{ μm}$$

10.9 A single mode step index SiO_2 fiber has the following specifications:

$$n_2 - n_1 = 0.012$$
$$a = 2.5 \text{ μm}$$
$$n_2 = 1.47$$

Using the material dispersion curve for SiO_2 shown in Figure 10.18 and the description for waveguide dispersion given in Eq. 10.54, compute the wavelength at which the sum of the two dispersions equals zero.

REFERENCES

1. Abramowitz, M., and Stegun, I. A. *Handbook of Mathematical Functions.* New York: Dover, 1965.

2. Barnoski, M. K. *Introduction to Integrated Optics.* New York: Plenum, 1974.

3. Cherin, A. H. *An Introduction to Optical Fibers.* New York: McGraw-Hill, 1983.

4. Fleming, J. W. "Material dispersion in light-guide glasses." *Electronics Letters* 14 (1978): 326–328.

5. Glenn, A. B., and Keiser, G. E. "A perspective of lightwave technology." *IEEE Communications* (1985): 8–9.

6. Gloge, D. "Dispersion in weakly guiding fibers." *Applied Optics* 10 (1971): 2442–2445.

7. Gloge, D. "Weakly guiding fibers." *Applied Optics* 10 (1971): 2252–2258.

8. Henry, P. S. "Introduction to lightwave transmission." *IEEE Communications* 23 (1985): 12–16.

9. Keck, D. B. "Single-mode fibers outperform multimode cables." *IEEE Spectrum* (March 1983): 30–37.

10. Keiser, G. *Optical Fiber Communications.* New York: McGraw-Hill, 1983.

11. Kong, J. A. *Theory of Electromagnetic Waves.* New York: Wiley, 1975.

12. Lemrow, C. M., and Bhagavatula, V. A. "Advanced fiber designs." *Laser Focus* (March 1985): 82–92.

13. Moriyama, T., *et al.* "Ultimately low OH content V.A.D. optical fibers." *Electronics Letters* 16 (1980): 699–700.

14. Okoshi, T., *Optical Fibers.* New York: Academic, 1982.

15. Osanai, H., *et al.* "Effects of dopants on transmission loss of low OH content optical fibers." *Electronics Letters* 12 (1976): 549–550.

16. Snyder, A. W. "Understanding monomode optical fibers." *Proceedings of the IEEE* 69 (1981): 6–13.

17. Snyder, A. W., White, I., and Mitchell, D. J. "Radiation from bent optical waveguides." *Electronics Letters* 11 (1975): 332–333.

18. Suematsu, Y., and Iga, K. *Introduction to Optical Fiber Communications.* New York: Wiley, 1982.

APPENDIX 1

The dashed line shown in Figure 4.21b indicates the location of the electric field maxima or minima for any quasi-even or quasi-odd cover mode in the continuum limit. Requiring the number of cycles between this line and upper and lower plates to be equal implies that

$$k_{1x}l_1 + k_{2x}(d/2 + \Delta) = k_{3x}l_3 + k_{2x}(d/2 - \Delta)$$

or

$$\psi_1 - \psi_3 = -2k_{2x}\Delta \qquad (A1.1)$$

From Eqs. 4.51 and 4.52, for each quasi-even mode, ψ_1 and ψ_3 obey the relations

$$\tan\psi_1 = A_{12}\cot(\alpha - \psi) \qquad (A1.2)$$

$$\tan\psi_3 = A_{32}\cot(\alpha + \psi) \qquad (A1.3)$$

where

$$A_{12} = (\mu_2 k_{1x})/(\mu_1 k_{2x}) \qquad (A1.4)$$

$$A_{32} = (\mu_2 k_{3x})/(\mu_3 k_{2x}) \qquad (A1.5)$$

$$\alpha = k_{2x}d/2 \qquad (A1.6)$$

Corresponding to each quasi-even mode with phase ψ there exists a quasi-odd mode with phase $\psi + \pi/2$. From expressions A1.2 and A1.3 this mode must have corresponding phases ψ_1' and ψ_3' given by

$$\tan\psi_1' = A_{12}\cot\left(\alpha - \psi - \frac{\pi}{2}\right) = -A_{12}\tan(\alpha - \psi) \qquad (A1.7)$$

$$\tan\psi_3' = A_{32}\cot\left(\alpha + \psi + \frac{\pi}{2}\right) = -A_{32}\tan(\alpha + \psi) \qquad (A1.8)$$

Because the quasi-odd mode is centered about the same line as its corresponding quasi-even mode we require

$$\psi_1' - \psi_3' = -2k_{2x}\Delta \qquad (A1.9)$$

Equating A1.1 and A1.9 therefore yields the requirement

$$\psi_1 - \psi_3 = \psi_1' - \psi_3' \qquad \text{(A1.10)}$$

Substituting Eqs. A1.2, A1.3, A1.7 and A1.8 into Eq. A1.10 yields

$$
\begin{aligned}
\tan^{-1}[A_{32} \cot(\alpha + \psi)] &- \tan^{-1}[A_{12} \cot(\alpha - \psi)] \\
= \tan^{-1}[A_{12} \tan(\alpha - \psi)] &- \tan^{-1}[A_{32} \tan(\alpha + \psi)]
\end{aligned}
\qquad \text{(A1.11)}
$$

Taking the tangent of both sides of A1.11 and making use of the identity

$$\tan(a + b) = \frac{\tan a + \tan b}{1 - \tan a \tan b}$$

we find after some algebraic manipulation that

$$A_{32}(1 - A_{12}^2) \sin[2(\alpha - \psi)] = A_{12}(1 - A_{32}^2) \sin[2(\alpha + \psi)]$$

Expanding the sine and cosine terms then yields

$$\tan 2\psi = \tan(2\alpha) \left(\frac{1 - \beta}{1 + \beta} \right)$$

with

$$\beta = \frac{A_{12} \, (1 - A_{32}^2)}{A_{32} \, (1 - A_{12}^2)}$$

Making use of the definition of α, A_{12}, A_{32} from above yields Eq. 4.54.

APPENDIX 2

We show that expressions of the form

$$x = F(x) \qquad (A2.1)$$

can be solved for x by iteration, provided that

$$|\partial F(x)/\partial x| < 1$$

Proof

Let us define x_0 as the true solution to Eq. A2.1. Further, let x_1 be the initial guess for the solution. Our next guess is obtained from (A2.1) as

$$x_2 = F(x_1) \qquad (A2.2)$$

Provided that our initial guess is not too far off from the correct value, x_0, then $F(x_1)$ can be expanded to good approximation in a first-order Taylor series as

$$F(x_1) \cong F(x_0) + (x_1 - x_0) F'(x_0)$$

But by definition, $F(x_0) = x_0$; thus,

$$F(x_1) \cong x_0 + (x_1 - x_0) F'(x_0) \qquad (A2.3)$$

and therefore from (A2.2)

$$x_2 = x_0 + (x_1 - x_0) F'(x_0) \qquad (A2.4)$$

Our next guess, x_3, is similarly given by

$$x_3 = F(x_2)$$
$$= x_0 + (x_2 - x_0) F'(x_0)$$

which upon substitution of Eq. A2.4 yields

$$x_3 = x_0 + (x_1 - x_0) [F'(x_0)]^2$$

Extending to the qth guess, therefore yields

$$x_q = x_0 + (x_1 - x_0) [F'(x_0)]^{q-1}$$

In the limit that $q \rightarrow \infty$ and provided that $|F'(x_0)| < 1$ we have

$$\lim_{q \to \infty} x_q = x_0$$

which is the desired result.

APPENDIX 3

Consider a cylindrical volume of differential length Δz. Let ϕ and ψ be any two arbitrary functions. Further, define the vector quantity $\mathbf{R}$ as

$$\mathbf{R} = \psi \nabla \phi$$

Taking the divergence of $\mathbf{R}$ gives

$$\nabla \cdot (\psi \nabla \phi) = \psi \nabla^2 \phi + \nabla \psi \cdot \nabla \phi$$

Integrating over the volume V and making use of Gauss' theorem then yields

$$\oiint_S ds(\psi \nabla \phi) \cdot \hat{n} = \iiint_V dv(\psi \nabla^2 \phi + \nabla \psi \cdot \nabla \phi) \qquad \cdot (\text{A3.1})$$

where the surface S bounds V and $\hat{n}$ is an outward normal to this surface. For Δz sufficiently small, all field quantities are uniform with respect to z and the integration over this variable may be replaced with the multiplicative constant Δz.

Equation A3.1 then reduces to

$$\oint_C dl(\psi \nabla \phi) \cdot \hat{n} = \iint_A dA(\psi \nabla^2 \phi + \nabla \psi \cdot \nabla \phi) \qquad (\text{A3.2})$$

where C is the closed line path bounding the surface A. If the radius of the surface A is allowed to approach infinity and provided that ψ decreases exponentially with r, the left side of Eq. A3.2 is zero. Since for our purposes ψ and ϕ are independent of z, the ∇ operator can be replaced with its transverse component portion ∇_t and the desired result, Eq. (10.4), is obtained.

Index

Acoustic energy, 264
Acoustic wave, 247
  surface, 8, 258, 261
  velocity, 247, 271
Acousto-optic beam deflector:
  bandwidth, 253–254
  bulk wave, 251–254
  guided wave, 257–269
Acousto-optic effect, 247
Acousto-optic spectrum analyzer, 254–257, 267–269
Angle:
  Brewster, 74
  of incidence, 53
  of reflection, 53
  of transmission, 53
Asymmetric slab waveguide, 86–109
  cover radiation modes, 102–108
  dispersion relation, 92–94
  effective index, 92
  effective waveguide thickness, 95–96
  guidance condition, 89, 90–92
  guided modes, 87–89
  substrate radiation modes, 102–103, 108–109
  TE modes, 87–89
  TM modes, 89
Asymmetry measure, 92, 94
Attenuation:
  from bending losses, 116, 135–142, 299–301
  in ion-migration waveguides, 177
  from material absorption, 298–299
  in optical fibers, 298–301
  in polymer-film waveguides, 176
  in proton-exchange waveguides, 179
  from scattering, 299
  in sputtered-film waveguides, 175
  of various transmission media, 275

Background field, 14
Backward coupling, 231–243
Backward traveling mode, 211

Bandwidth:
  of acoustic transducer, 270
  of acousto-optic beam deflector, 253–254
  of grating filter, 241–243
  optical, 1
Beam deflection:
  acousto-optic, 251–254, 257–269
  electro-optic, 5, 257–258
Beam spread, see Diffraction
Bell, Alexander Graham, 1
Bending loss, see Radiation bending losses
Bessel functions, 289–290, 292–294
  asymptotic values, 290
Bouncing beam concept, 77, 84, 90–92, 96, 98, 102–104, 115, 312
Boundary conditions, at dielectric interface, 51–52
Bragg angle, 250
Bragg condition, 8, 247–251, 263
Bragg deflection efficiency, 262, 265–266
Bragg scattering, 247–269
Brewster angle, 74
Bulk wave:
  acousto-optic spectrum analyzer, 254–257
  beam deflector, 251–254
  modulator, 247
Buried channel waveguide fabrication, 178, 185–186, 198

Carrier frequency, 26, 302
Characteristic impedance of free space, 23
Charge density, 13
Chemical vapor deposition (CVD), 189–192
Circular polarization, 32–33
Collimation distance, 38
Communication, optical, 1
Compressional wave, 247
Conduction current, 39
Conservation of power:
  for grating reflector, 237
  Poynting, 41, 45
  reflected and transmitted, 65
Constant phase planes, 20–21

Constitutive relations, 14
Continuity conditions, 51–52
Continuity equation, 13
Core, dielectric, 77, 276
Coupled mode equations, 218–243
  backward coupling, 231–243
  Bragg deflection, 258–263
  forward coupling, 224–231
Coupled waveguides, 220–224
Coupling:
  to backward waveguide mode, 231–243
  Bragg, 258–263
  by evanescent fields, 147
  to forward waveguide mode, 224–231
  length, 164, 226–227
  prism, 147–168
Coupling coefficient:
  backward coupling, 233–234, 239–240
  between guides for TE modes, 229–231
  Bragg scattering, 261, 266
  forward coupling, 223–224
  TE mode in grating, 239–240
Cover radiation modes, 103–108
Critical angle, 56–57, 63–64, 66–67, 77, 84,
  103–104, 147, 152, 314
Current, 13
  conduction, 39
  source, 39
Cutoff:
  for graded-index waveguide, 121
  for multimode optical fiber, 295–296
  for slab waveguide, 82, 85, 91–92
Cutoff frequency:
  for graded-index waveguide, 121
  for slave waveguide, 83
CVD (chemical vapor deposition), 189–192
Cylindrical dielectric waveguide, 209

Delay:
  group, 28
  phase, 28
Depressed cladding optical fiber, 311
Dielectric core, 77, 276
Dielectric grating, *see* Grating
Dielectric waveguide, *see* Waveguide(s)
Diffraction, 36–38, 137, 251, 255
Diffusion coefficient, 180, 201
Directional coupler, 6, 224–231
Dispersion:
  of Gaussian pulse, 301–307
  intermodal, 312–314
  material, 307–309

waveguide, 309–312
Dispersion relation, 19, 29
  for asymmetric slab waveguide, 92–94, 100–
    102
  for embedded-strip waveguide, 132
  for Gaussian-index waveguide, 315
  for ionized plasma, 30
  for optical fiber, 295
  for symmetric slab waveguide, 85–86
Divergence theorem, 41
Dose, implant, 181
Dry etching for waveguides, 195
Duality, 33–35
  for guided wave modes, 89
  for radiation modes, 108–109
  for reflection coefficients, 61

Effective figure of merit, 266, 272
Effective index, 92
Effective index method, 115, 128–135
Effective refractive index, 92, 120–126, 166, 286
Effective waveguide thickness, 95–96, 110, 163
Electric displacement, 13
Electric field, 13
Electromagnetic field(s), 13
  background, 14
  time-harmonic, 15
Electro-optic:
  guided-wave beam deflection, 257–269
  effect, 6, 247, 257–258
  modulator, 5
  switch, 6, 229
Elliptical polarization, 33
Embedded strip waveguide, 113–115, 128–135,
  198
  effective index method for, 128–135
Empedocles, 1
Energy, acoustic, 264
Energy density, 41, 46
Euclid, 1
Evanescent field coupling, 147
Evanescent fields:
  for optical fiber, 292
  for plane wave, 57–58
  for prism coupler, 147
  for slab waveguide, 78
  for totally internally reflected wave, 66–67

Fiber, *see* Optical fiber
Fields, *see* Electromagnetic field(s)
Film characterization, 166–168

Film thickness measurement, 166–168
Filter, grating, 7, 241–243
Flame hydrolysis for optical fibers, 199–200
Focal length, 255
Focal plane, 255
Forward coupling, 224–231
Fourier spectrum, 9–10, 27, 254, 304
Free carrier index waveguide, 186
Free space:
  characteristic impedance, 23
  permeability, 14
  permittivity, 14
  phase velocity, 24–25
Fresnel lens, 267–269
Frequency of cutoff:
  graded-index waveguide, 121
  slab waveguide, 83
Frustrated total internal reflection, 147

GaAlAs waveguide, 185–186
GaInAsP waveguide, 185–186
Galileo, 1
Gaussian refractive index profile, 315
Gauss' law, 41
Geodesic lens, 267–269
Geometric series, 150
Glass, 176
  OH$^-$ contamination, 200
  for optical fibers, 199–200
  refractive index of, 175
  soda-lime, 177
Goos-Haenchen shift, 67–73, 95–96, 162
Graded-index fiber, 277, 315
  intermodal dispersion, 312–314
Graded-index waveguide, 113, 116–127, 142–144, 176, 277, 315
Grating:
  for Bragg deflection, 247–250, 257–258
  filter, 241–243
  reflection coefficient, 7, 231–240
  for spectrum analyzer, 7–9, 254–257, 267
Group delay, 28
Group velocity, 26–31, 306
Guidance condition:
  asymmetric slab waveguide, 89, 90–92
  highly asymmetric slab waveguide, 109
  multimode step-index fiber guide, 294
  radiation modes on slab waveguide, 98–100
  ray interpretation, 90–92
  symmetric slab waveguide, 80–82
Guided wave beam deflection, 257–269

High frequency waveguide limit, 83–86
Hooke's law, 263

Impedance of free space, 23
Implant, *see* Ion-implanted waveguide
Information capacity:
  of optical fibers, 1, 301
  of various transmission media, 275
Inhomogeneous plane wave, 57–58
Instantaneous energy density, 41
Interferometer, 5
Intermodal dispersion, 312–314
Ion-implanted waveguide, 181–185, 198
Ion-migration waveguide, 176–178, 198
Isotropic medium, 14
Iterative solutions, 166–168

K-space diagrams, 54, 56–57, 153, 161–162, 250, 252
K-vector, 20–21

Laser, 1–3, 267
Lateral beam shift in waveguides, 67–73, 95–96, 162
Lattice matching, 186
Leaky modes, 85
LED (light-emitting diode), 277
Length, coupling:
  prism, 164
  waveguides, 226–227
Lens:
  planar, 8, 267–269
  properties of, 255
Light-emitting diode (LED), 277
Lightwave communication, 1
LiNbO$_3$:
  electro-optic switch, 229
  waveguide by ion implantation, 185
  waveguide by LPE, 188
  waveguide by metal in-diffusion, 180–181
  waveguide by proton exchange, 178–180
Linear polarization, 32
Linearly polarized modes, 294–298
LiTaO$_3$ waveguide:
  for beam deflection, 257
  by ion implantation, 184–185
  by LPE, 188
  by metal in-diffusion, 180–181
  by proton exchange, 178–180
Liquid phase epitaxial waveguide (LPE), 186–189
Longitudinal acoustic wave, 247

Lorentz reciprocity theorem, 212–215, 260–261
  for cylindrical waveguides, 215
Low frequency limit, 82–86
LPE (liquid phase epitaxy), 186–189
Luneburg lens, 267

Mach-Zehnder interferometer, 5
Magnetic displacement, 13
Magnetic field, 13
Mask and etch technique, 194–196
Material dispersion, 307–309
  of pure silica, 308
Maxwell's equations, 13
  longitudinal and transverse components, 209–211
  for plane waves, 22–23
  time-dependent form, 13
  time-harmonic form, 14–17
MBE (molecular beam epitaxy), 192–194
Metal in-diffusion for waveguides, 180–181
Metal organic chemical vapor deposition (MOCVD), 191–192
Minimum resolvable spot size, 255
MOCVD (metal organic chemical vapor deposition), 191–192
Mode orthogonality, 216–218
Modes, *see* Waveguide mode
Modified Bessel function, 292
Modulator:
  acousto-optic, 247
  electro-optic, 6
Molecular beam epitaxy (MBE) waveguide fabrication, 192–194
Monomode optical fiber:
  effective refractive index, 286
  fields, 283
  radiation bending loss, 299–301
  refractive index profile, 277
  spatial light intensity, 287–288
  stationary form for propagation constant, 279–287
  wave equation, 278–279
  waveguide dispersion, 309–312
Multimode optical fiber:
  boundary conditions, 293–294
  cutoff condition, 295–296
  dispersion relation, 295
  fields, 291–293
  guidance condition, 294
  refractive index profile, 277
  spatial intensity, 297–298
  wave equation, 289

Newton's law, 271
Normalized field amplitude for waveguide modes, 218
Normalized frequency, 92
Normalized thickness parameter, 286
Numerical aperture, 314

OH$^-$ ion contamination in glass, 200
Optical bandwidth, 1
Optical beam:
  diffraction, 36–38
  plane wave spectrum, 36–38, 67–69
Optical communication, 1, 275
Optical fiber:
  absorption, 298–299
  attenuation, 298–301
  cladding, 276–277
  depressed cladding, 311
  fabricated by flame hydrolysis, 199–200
  frequency transfer function, 304
  geometries, 277
  graded-index, 277, 315
  information data rate, 275
  loss mechanisms, 298–301
  materials, 276
  monomode, 277–288, 300–301, 309–312, 315
  multimode, 277, 288–298
  radiation bending losses, 299–301
  scattering, 299
  segmented profile, 311
  step-index, 277, 283–298
  triangular profile, 311
  W profile, 311
Optical interferometer, 5
Optical modulator, 5
Optical switch, 6, 229
Optical switching network, 6, 229
Optical waveguide, *see* Waveguide(s)
Orthogonality of modes, 216–218
Outside vapor deposition for optical fibers, 199–200

Parabolic index waveguide, 122–127, 142
Paraxial modes, 126–128, 279, 314
Pencil beams, *see* Ray trajectory
Permeability, of free space, 14
Permittivity, of free space, 14
Perturbed waveguide, 212, 220–224, 231–233, 258–259
Phase delay, 28
Phase matching, 52–57
  forward mode-coupling, 226–229

grating reflector, 237–243, 250
prism coupler, 160–162
Phase shift:
    due to propagation, 90
    due to TIR, 62–64, 66–67
Phase velocity, 24–25, 29
Phasor notation, 15
Photodetector, *see* Photodiode
Photodiode, 3
Photoelastic constant, 271
Photoelastic effect, 263–265, 271–272
Photophone, 1
Piezoelectric effect, 258
Piezoelectric material, 8
Piezoelectric transducer, 247
Plane of incidence, 53
Plane wave:
    boundary conditions at an interface, 53
    characteristic impedance, 23
    dispersion, 29
    dispersion relation, 19
    evanescent or inhomogeneous, 57–58
    fields, 22–24
    k-vector, 20–21
    group velocity, 26–29
    Maxwell's equations for, 22–23
    phase matching at interface, 52–58
    phase velocity, 24–25, 29
    polarization, 31–33
    propagation through plasma, 30
    reflection and transmission coefficient, 58–67
    reflectivity and transmissivity, 64–66
    spectrum, 36–38, 67–69, 137–139
    standing, 66–67
    superposition for guided modes, 84
    TE, 58
    TM, 61
    total internal reflection for, 63–67
    wave vector, 20–21
Plasma dispersion diagram, 30
Plasma frequency, 30
Plasma polymerization, 174–176
Polarization:
    circular, 32–33
    elliptical, 33
    linear, 31–32
Polymer film waveguide, 174–176
Power:
    coupled by a prism, 154–166
    density, 39, 45
    dissipated, 41, 46
    flow of, 41, 45
    generated, 41, 46

for guided mode, 94–96
for incident TE wave, 64
interchange between coupled guides, 226–229
Poynting, 39–47
reactive, 46
reflected by a grating, 236–239
reflected TE wave, 64
time-average, 45
transmitted TE wave, 64
Poynting's theorem,
    complex, 44–48
    time-dependent, 39–43
Poynting vector, 41
    for TE waves at dielectric interface, 64
Prism coupler, 147–168
Propagation angle for guided modes, 90–92, 96,
    115, 312
Propagation constant:
    for backward traveling modes, 211
    for guided modes, 78
    for optical fibers, 283
    stationary formulation, 279–289
Propagation direction, 20–22
Propagation loss, *see* Attenuation
Proton-exchange waveguide, 178–180
Pulse envelope, 26
Pulse spreading, 302–307

Radiation bending losses, 116, 135–142, 299–301
Radiation modes:
    asymmetric slab waveguide, 102–109
    cover, 102–108
    dispersion relation, 100–102
    guidance condition, 98–100
    substrate, 102–103, 108–109
    symmetric guide, 97–102
Raised channel waveguide fabrication, 194–198
Range, 182
Range straggling, 182
Ray summation technique, 148–152, 154–166
Ray trajectory:
    in inhomogeneous medium, 116–118
    for parabolic index guide, 126–127
    for slab waveguide, 77, 84, 90–92, 96–98,
        102–104, 312
Reactive power, 46
Reactive sputter etching, 195–196
Reflected power, 64
Reflected wave, 52–54
Reflection coefficient:
    for grating, 231–240
    for TE plane wave, 58–67

Reflection coefficient (*Continued*)
  for TM plane wave, 61–65
  at turning point, 119–120
Reflectivity:
  at dielectric interface, 64–66
  grating, 231–240
Reflector, grating, 7, 231–240
Refractive index, 83
  polymer film waveguides, 176
  sputter-film waveguides, 175
Rib waveguide, 113–115, 134–135
  fabrication, 194–198

SAW, *see* Surface acoustic wave
Scattering loss, 299
Segmented profile optical fiber, 311
Sellmeier equation, 203
Semiconducting waveguide:
  GaAlAs, 185–186
  GaInAsP, 185–186
Semiconductor laser, 1–3, 267
Shadow mask waveguide fabrication, 196–197
Signal distortion, 301–314
Silica material dispersion, 308
Slab waveguide, *see* Asymmetric slab waveguide;
  Symmetric slab waveguide
Snell's law, 54
Soda-lime glass, 177
Source current, 39
Source free medium, 18
Spatial light intensity:
  step index monomode fibers, 287–288
  step index multimode fibers, 297–298
Spectrum analyzer, 7–9
  acousto-optic, 254–257, 267–269
  frequency resolution, 256
  integrated optics, 267–269
Spectrum of plane waves, 36–38, 67–69, 137–138
Sputtering, 172–173
Standing waves, 66–67
Stationary formulation, 279–287
Step index fiber:
  monomode, 277–288
  multimode, 288–298
  see also Monomode optical fiber; Multimode
    optical fiber
Sticking coefficient, 172
Stored acoustic energy, 264
Stored electrical energy, 41–43, 46–47, 66
Stored magnetic energy, 41–43, 46–47, 66
Strain, 263–264
Stress, 247, 263
Strip-loaded waveguide(s), 113–115, 135
  fabrication, 194–198

Strip waveguide, 113–115
  fabrication, 194–198
Substrate radiation modes, 102–103, 108–109
Superposition of plane waves, 36–38, 67–69, 137–138
Surface acoustic wave, 8, 258, 261
  field distribution, 272
  transducer, 8
  transducer bandwidth, 270
  transducer power, 270
Switch, electro-optic, 6, 229
Switching network, 6, 229
Symmetric slab waveguide:
  coupling coefficient between two guides, 229–231
  coupling coefficient for grating, 239–240
  cutoff condition, 82
  cutoff frequency, 83
  dispersion relation, 85–86
  guidance condition, 80–82
  guided mode power, 94–96
  high frequency limit, 83–86
  low frequency limit, 82–86
  propagation constant, 82
  radiation bending losses, 135–142
  radiation modes, 97–102
  ray trajectory, 77, 84
  TE guided mode solutions, 79–80

TE plane wave, *see* Plane wave
TE radiation modes:
  boundary conditions, 98, 104
  dispersion relation, 100–102
  fields, 102, 108–109
  guidance condition, 99–102
  ray bounce angles, 97–98, 102–104
TE waveguide modes:
  boundary conditions, 79–80, 87–88
  dispersion relation, 85–86, 92–94
  effective waveguide thickness, 95–96
  fields, 79–80, 82–83, 87–89
  guidance condition, 80–82, 89, 90–92
  Poynting power, 94–96
Three-dimensional waveguides, 113–115, 127–135
  fabrication, 194–198
Time-average power, 45
Time-harmonic Maxwell's equations, 14–17
Time-reversal symmetry, 211–212
TM plane wave, *see* Plane wave
TM radiation modes, 108–109
TM waveguide modes:
  dispersion relation, 92–94
  effective waveguide thickness, 95–96
  fields, 89

guidance condition, 89
Total internal reflection, 4, 63–67
Transducer, SAW, 8, 270
Transmission coefficient:
    across dielectric interface, 58–67
    across two dielectric interfaces, 147–154
Transmission line:
    data rate, 275
    propagation loss, 275
Transmissivity:
    at dielectric interface, 64–65
    for prism coupler, 152–154
Transmitted power, 64
Transverse electric, *see* TE plane wave; TE
    radiation modes, TE waveguide modes
Transverse magnetic, *see* TM plane wave; TM
    radiation modes; TM waveguide modes
Triangular profile optical fiber, 311
Tunneling, 152
Turning points, 117

Vapor phase epitaxy (VPE), *see* CVD (chemical
    vapor deposition); MOCVD (metal organic
    chemical vapor deposition)
Variational method, 284
Velocity of an acoustic wave, 247, 271
Velocity of light, 25
    group, 26–31, 306
    phase, 24–25, 29
VPE (vapor phase epitaxy), *see* CVD (chemical
    vapor deposition); MOCVD (metal organic
    chemical vapor deposition)

Wave equation, 18–19
    in cylindrical coordinates, 289
    for monomode fibers, 278–279
Waveguide(s),
    asymmetric slab, *see* Asymmetric slab
        waveguide
    bent, 135–142, 299–301
    cylindrical, 209
    dispersion, 309–312
    effective thickness, 95–96, 110, 163
    embedded-strip, 113–115, 128–135, 198
    excitation with prism, 154–166
    free carrier index, 186
    GaInAsP, 185–186
    GaAlAs, 185–186
    graded-index, 113, 116–127, 142–144, 176,
        277, 315

LiNbO$_3$, 178–181, 185, 188, 229
LiTaO$_3$, 178–181, 184–185, 188, 257
    parabolic index, 122–127, 142
    perturbation, 212, 220–224, 231–233, 258–259
    rib or ridge, 113–115, 134–135
    semiconducting, 185–186
    strip, 113–115
    strip-loaded, 113–115, 135
    symmetric slab, *see* Symmetric slab waveguide
    three-dimensional, 113–115, 127–135
Waveguide fabrication:
    for buried channel waveguides, 178, 185–186,
        198
    CVD, 189–192
    dry etching, 195
    ion implantation, 181–185, 198
    ion-migration, 176–178, 198
    LPE, 186–189
    mask and etch, 194–196
    MBE, 192–194
    metal in-diffusion, 180–181
    MOCVD, 191–192
    plasma polymerization, 174–176
    proton exchange, 178–180
    for raised channel, 194–198
    shadow masking, 196–197
    sputter deposition, 171–174
    for three-dimensional waveguides, 194–198
    wet chemical etching, 196
Waveguide mode, 78, 209
    asymmetric slab, 86–94
    backward traveling, 211
    leaky, 85
    normalized fields, 218
    orthogonality relations, 216–218
    propagation angle, 77, 84, 90–92, 96–98, 102–
        104, 115, 312
    propagation constant, 78
    radiation, 97–109
    symmetric slab, 79–86
    paraxial, 127
Wave vector, 20–21
Wavenumber, 56
Weakly guiding fibers, *see* Monomode optical fiber
W profile optical fiber, 311
WKB method, 116–127

Young's modulus, 264

Z-reversal symmetry, 209–211